Parkinsons Recovery provides support, information and resources to individuals who have the symptoms of Parkinson's disease. Five Steps to Recovery provides a foundational approach for transforming thought forms held by persons with Parkinson's. Transforming thought forms is the key to maintaining a steady course on the road to recovery.

ISBN 978-0-9819767-2-3

Table of Contents

 Robert Rodgers, Ph.D.

Introduction

As I am writing this introduction, I am preparing to travel to Casper, Wyoming where I will attend my uncle's funeral. A former college president and Wyoming legislator, my uncle Gordon Ward was an authentic intellectual and a man of many curiosities and talents. As a child, I found myself mesmerized by the depth of his knowledge and curiosity. Gordon died several weeks ago from the complications of Parkinson's Disease.

I can state unequivocally that I have had my fill of Parkinson's Disease. My mother, also diagnosed with Parkinson's Disease, died 11 years ago from a stroke that was aggravated by the interactions of the many medications she was taking. Enough is enough.

I am now into my fourth year of research into the factors that cause the symptoms of Parkinsons and the therapies that give people relief from their symptoms. I take great pride in the recent publication of my book Pioneers of Recovery, which is a compilation of 14 interviews with people who are genuine, certified pioneers of recovery in my book. I have received a barrage of unsolicited rave reviews. People are taking action on one or another of the suggestions and finding they now feel a lot better. This news warms my heart.

I have also been chasing after the biggest questions of all: What needs to happen for a person to become symptom free? What did my uncle and mother need to know to avoid decades of suffering and trauma? I became convinced six months ago that there had to be a dark cloud of energy that was making it impossible for my mother and my uncle to get better.

What was responsible for their suffering, I asked myself? From the research I have done on Parkinson's, I knew it had to be more than a deficiency of dopamine which, in my opinion, is a simpleminded and terrible restrictive perspective.

Deborah Russell and I landed on the key after reading Mary Browne's wonderful book *Five Rules of Thought.* Her precious book is simply written, clear, succinct and dead right in every respect.

Deborah decided to do the exercises Mary recommends in her book with wonderful results. The ideas are palatable. The rules in Mary's book are well know and well documented by many other writers. Her gift has been to synthesize the ideas and present them coherently, simply and practically.

I personally worked with each of Mary's rules for myself, one by one. Results were promising. So, I began translating and expanding Mary's rules for people interested in recovering from Parkinson's. My terminology differs of course. I talk about steps, not rules. The material you will find in Five Steps to Recovery speaks to the very specific issues confronted by anyone with the symptoms of Parkinson's. I have recorded 34 meditations that explicate each step.

I am excited about this book because it has had a profound influence on my own life. I now meticulously monitor my own thoughts moment by moment. I honor and acknowledge the power of my own thoughts. My vision is to help millions of people who have the symptoms of

Parkinsons find significant relief from their symptoms. I cannot accomplish my vision with wrong thinking.

I honor the fate of my mother and uncle. Their experience with Parkinson's has been a true inspiration for my life's work. I know in my heart there can be a different fate for anyone who has been diagnosed with Parkinson's. Five Steps to Recovery presents the roadmap for how recovery can unfold for anyone.

I am deeply indebted to Deborah Russell, my partner in love and work, for all her guidance, insights and assistance on this project. My photographer son Jonathan Rodgers and Marcie Grim assisted with acquiring the beautiful photographic images that appear throughout the book.

Take the ideas seriously. Do the daily meditations. I predict your life will change. Mine has.

Robert Rodgers, Ph.D.
Parkinsons Recovery
Olympia, Washington
www.parkinsonsrecovery.com
Robert@parkinsonsrecovery.com

P.S. On the membership website, I post the interviews, meditations and radio programs that I do each week. Find out more about Parkinsons Recovery Membership at www.parkinsonsrecovery.org.

Step One: Focus

So, what are the five steps? I suspect that the most of you who reading this may be anticipating that I'm going to be giving you specific information about the first step you need to take in order to be able to jump start your recovery from the symptoms of Parkinson's.

 Robert Rodgers, Ph.D.

That is right. That's exactly what this chapter is about. Part of me also suspects that some of you are probably thinking I'm going to be saying something to the effect of,

> *"Oh, the first step is to be careful about what you put in your body. Watch what you put into your body."*

Or, perhaps some of you are thinking

> *"Oh, exercise is the most important first step."*

Well, guess what? I am not going to talk about any of those specific types of actions and modalities and therapies that I do a lot of writing about and that you see discussions about in the Parkinsons Recovery blogs and newsletters

Instead, I will talk about most important steps that are required if in fact you want to launch a program to recover. By this I mean, steps you can take to see a significant relief from the symptoms of Parkinson's or to become literally symptom free.

- *What is the secret?*
- *What is the first step?*

Ah, you want me to be clear here? Let me give you the experience when someone waffles – when they do not directly answer themselves, when their thinking is fuzzy and unclear.

Five Steps to Recovery

I received a rather interesting correspondence a few days ago from an individual who wrote he found the types of things I write about intriguing and interesting, but was wondering "what's the hidden agenda"?

> *"I keep waiting for the product that you're going to offer that's going to solve everybody's problem. So, what is it? What's the secret? What's the hidden agenda here?*

I must admit I was surprised by the question. I did not see it coming. I responded with:

> *"What you see is what you get.*
>
> *"No there really isn't any secret product or a hidden agenda. No, I'm not going to be a coming out after you've been with me for several weeks or months or years and say:*
>
> *"Hey, now that you're a veteran member at Parkinsons Recovery I am going to give you the secret but, of course, you have to first fork over $5000 American dollars or maybe $10,000 American dollars or why not $50,000 American dollars? Once I receive the bank transfer I'll go ahead and tell you the grand secret.*

Well of course the truth is (as you can well imagine)

> **There is no grand secret.**
> **There is no secret cure**

I do have to confess there is an essential, fundamental secret to being able to launch your recovery program successfully so that you begin to get relief from the symptoms of Parkinson's. There is no doubt about that.

There is a very important step that you can actually do to get relief from your symptoms. What is it?

The answer is:

You need to have clarity of thought.

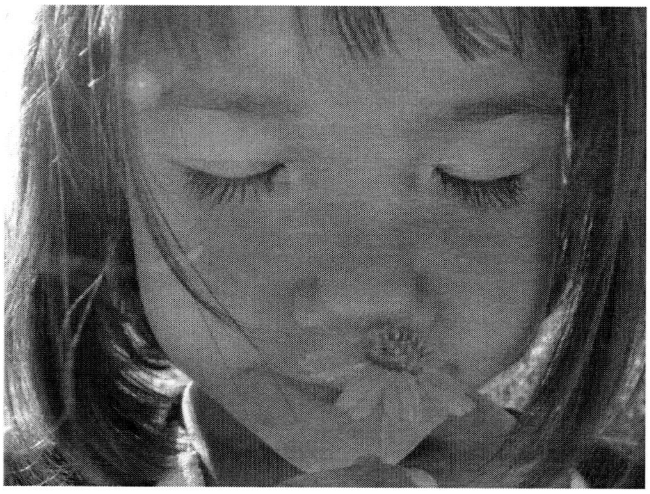

Thoughts are the most powerful force in the universe. With clarity of thought, you can manifest anything your heart desires.

So you ask,

> *"Thought is the most powerful force in the universe. Really? Give me a break. Just because I think something doesn't mean that it's going to actually happen."*

Of course that's true but step back for a moment and consider the power of clear thinking.

Before life existed there was thought.

Thought is responsible for creating life in the first place (with a little help from divine intervention). Many people ponder year after year writing book after book trying to answer the question;
What existed before the universe existed, before the physical manifestation of

- *the planets*
- *the galaxies*
- *the stars*

and all that we now know exists in physical form? I don't have to write books or ponder that question for decades. I know the answer.

Thought existed before the universe was physically manifested. Clarity of thought is necessary to manifest anything.

Thought is potentially more powerful than oil believe it or not.

Thoughts are more powerful than

- *Coal*
- *Electricity*
- *Nuclear power*
- *The strongest of the strongest men and women in the world*

Thought is the most powerful concept that exists in the world.

Think about what happens when you actually set an intention to get well. You might just do that right now. Get out a piece of paper and pencil or pen. Answer the following question.

I'm going to ask you to just answer it, whatever comes off the top of your head, write down your answer.

"What would you like to see happen with regard to the symptoms that you now have? What would you like to see happen?

Write whatever comes to mind first whatever works right them down

- *No censorship*
- *No screening*
- *No judgment*

Write down on your piece of paper the thought that just popped into your mind.

Let me now talk about the meaning of the words that you just committed to paper:

How specific, concise and concrete is the intention that you have just set for yourself? Did you simply write,

> *"I want to feel better."?*

If so, that can encompass a wide range of outcomes

- *In any moment of the day we can feel better*
- *It's possible*
- *In any moment of the day we can feel happiness or joy*

I suspect that if that is your intention, within the next hour or perhaps two hours you will fulfill your wish. You will get what it is that you declared was important for you to have.

Consider an alternative possibility that you might have committed to paper. Perhaps you wrote,

My intention is for my symptoms not to get worse.

Actually, many people write just that on their clinic intake forms in response to the question

"What you want to get out of this session today"

A frequent response is:

"I just don't want to feel any worse"

As it turns out, people who state this intention typically manifest this intention. Their symptoms may shift here and there, but overall they will feel about as well off today as they will for five or six or 10 years from now. They manifest precisely what they thought they needed to manifest.

The clarity of the intention clearly makes a huge difference.

- *The more specific you can be,*
- *The more concrete that you can be,*
- *The more likely it is that you will be able to manifest your fondest intention quickly.*

The point of this all is that messy and chaotic thoughts don't get you very far because they lack the energy that is required to manifest those thoughts into physical form. What is required is

Specificity
Focus
Discipline

on a very concrete goal. Even with the best of intentions, even with specificity, focus and discipline about a concrete goal, challenges with clear thinking can obstruct the process of manifestation.

Consider an example. Take a person who has written the following in response to the question that I ask previously:

I want to take my daily walks with poise and grace.

Granted, this intention is indeed very specific and focused. There's nothing chaotic about the intention. It has the force of discipline.

Perhaps this person is you in terms of what you'd like to see with your recovery program. What is likely to happen now in the following minutes or next hours or days as you begin to see that intention unfold?

Well if you are like most people, it is likely that thoughts which are,

Messy
Negative
Vague
Chaotic
Undisciplined

Creep into your head. Permit me to ramble a bit and give you a flavor for the type of thoughts that are likely to emerge inside you.

- *There is a problem with my walking stick is just not working as well as I would like it to work*
- *I'm really disappointed with my progress*
- *I have no one to walk with on a regular basis so I might as well forget it*
- *I might fall. Whether I have my walker or have my walking stick I might actually fall.*
- *I'll be terribly embarrassed. Everyone will see me.*
- *It is too risky to walk in the cold and snow. After all, it is winter at least in this hemisphere of the universe*

We are talking here about hours and hours and days and days of thoughts that feed on one another. Thoughts tend to become more personal, more judgmental.

> *"Well, you know what. I am physically incompetent. I was a strong person years ago but that's now gone. It will never happen again. My youth is gone forever."*
>
> *"I am basically incompetent, slothful and basically I'm at the end of my ropes. The fact is I hate my body. That's all there is to it. My body is just not what needs to be. I don't even want to hang around and wait for the end of it all."*

Do you feel how defuse the focus is here?

The positive intention

I want to take my daily walks with poise and grace.

Does not and cannot have enough significant energy to manifest itself in physical form.

There is a discipline that is required in any successful recovery program The discipline is to formulate a very specific intention, just one intention at a time, and to focus on that intention with clarity so that there is no confusion about what it is that you would like to see happen.

- *No negative thoughts*
- *No chaotic thinking*
- *No messy thinking*

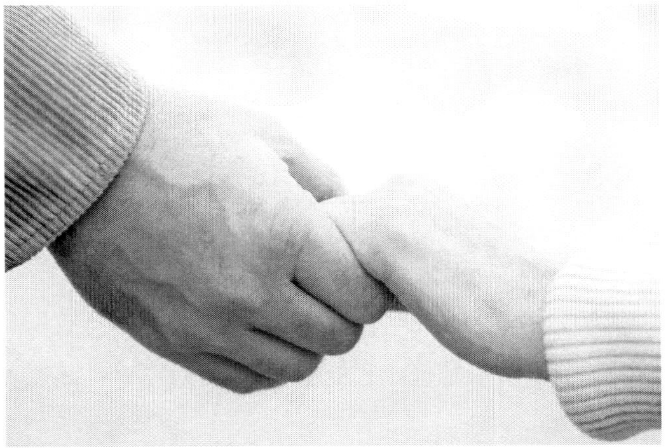

Clarity of focus on a clear intention sets into motion a cascade of possibilities that will take you to the next step. It may not be the case that within a week you will be able to satisfy the intention
But,

- *If you stick with it and*
- *If you focus single-handedly on that intention with your thought forms without adding all of the negative chaotic caveats*

you will see significant progress toward being able to fulfill and manifest what your heart desires.

Meditations

Focus Overview

What do you want? You are listening to this meditation because you want something. What is it?

Do you want one thing today and something else tomorrow? Is there one symptom that emerges in your face screaming

"Look at me I want to be noticed today."

That wish then vanishes in a flash by the end of the day as another symptom smacks you in your face. Do you want to address the first symptom or the second symptom or both?

It can be oh so frustrating. Symptoms shift moment to moment. Feelings are in perpetual flux. What we feel one moment shifts to something entirely different the next moment. Thoughts percolate throughout every cell of our body second by second, not in trickles but bursting dams of thoughts with water that overflows us and tends to drown us in our misery.

Five Steps to Recovery

A key to recovery is to make a decision. What is it that you want?

- *One thing.*
- *One wish.*
- *One desire.*

What is it that your heart desires? As you answer this question to yourself you may have already thought of

- *two or*
- *three or*
- *eight or*
- *nine*

fond wishes that you would like a genie to fulfill. Two or three or eight or nine wishes are too many to manifest in physical form. Recovery from the symptoms of any disease – and in our case Parkinson's - is a function of your ability to maintain focus on a single want.

- *Not two wants*
- *Not eight wants*
- *Not nine.*
- *Only one want.*

So you see, the true challenge is to make a decision. Decisiveness has a high vibrational frequency.

Decisiveness makes it possible to materialize the want. Indecisiveness does just the reverse. If we want "a" and then "b" and then "c" and then "d" all in the matter of an hour, none of those wants will ever be realized or manifested. There is too much to do and too little energy to make any of that physically possible.

Recovery happens when focus is present. How can focus really be possible for anyone?

The answer is practice and discipline. Anyone can do it when they set the intention to make it so. Practicing makes all the difference in the world. It is the discipline of the mind that allows us to maintain focus.

Why is it that some people with the symptoms of Parkinson's seem to be leading normal lives, doing all that they would like to do for themselves and others? And, why is it that some people are not?

One answer is that the person who is able to lead a fulfilling and active and exciting life is the person who has learned the discipline of focus - of being able to hold on and maintain

- one want
- one wish
- every day
- day after day

until that want or wish is manifested in physical form. And when it is, there can then be a second choice, a second want, a second wish that you place at the top of the list of wants. With focus, that wish too will be manifested in physical form

- not within hours most likely
- not within days

unless you are lucky. But in time - with discipline and focus - that want or wish will my dear friends be materialized.

So, the trick of it all is to set an intention to learn the discipline of focus. Focus on one thought. When there is focus on one thought, one wish, one want, the thought and wish and want has the power, the energy and the ability to materialize.

Everything we see in the physical world was preceded by thought. Thoughts are the most powerful force in the universe

- *more powerful than nuclear power*
- *more powerful than electricity*
- *more powerful than coal power*
- *more powerful than all of the military might of all the armies of all countries and nations.*

Thoughts are the supreme power. When you focus on a single thought, no one can stop that from materializing.

- ***Focus.***
- ***Discipline.***
- ***Decisiveness.***

These are the traits necessary to launch a successful program of recovery.

Nothing here is keyed to any specific want or wish. There is nothing for per se that must percolate to the top of your want or wish list. You see, it matters not what that particular want or wish it is. What matters is you maintain your focus on that want or wish -

- without hesitation
- without being wishy-washy
- without pulling back

Determination.

Decisiveness.

That is the leadership you need to be able to make happen what your heart so desires.

What then is it that you want, really? Think about it. What now is important to you? The list may already be three or four or five piled high. Step back and see what floats to the top. That is the most important want or wish. It may be linked to how your body is feeling today. It may be linked to a specific symptom.

But it very may very well not be directly linked to anything connected to your physical body.

- It may have to do with a pleasure that you have given up for whatever reason.
- It may have to do with the gift you wanted to give to someone else that you have decided you do not have the energy for
- It may have to do with connecting with a loved one who has been absent from your life too long.
- It may have to do with getting out of the closet a
    - paintbrush or
    - a musical instrument or
    - a journal that you started years ago and forgot to continue entering your thoughts

What is your wish or want for now, for today? What is percolating up to the top for you?

This you see is where most people get stuck. They do not know what it is that they truly want or wish. They are wishy-washy. They are indecisive. What they say is what perhaps they think others want them to say. What they think is what perhaps they think others want them to think.

Check in with yourself, not with your best friend or spouse or child or grandchild or parent or grandparent. Check in with yourself.

- *This is not about what somebody else wants for you.*
- *This is not about fulfilling someone else's wish for you.*

This is about deciding what it is you want for yourself.

Open up the door to all possibilities. This may be connected to work or play or pleasure. Ask yourself,

- ***"What is it that I most want now, for me."?***

As you ponder the choices, what are the feelings? Where are you getting energy? With what choice does your heart begin to palpitate? With what choice do you want to stand up and race to do it?

Open up the door to all possibilities. What may give you a surge of energy is a choice you have not thought about for decades.

What is it that you truly loved to do as a child? How did you spend your time? Who did you love hanging out with, if anyone? What were you doing when you had a choice about what you could do?

Remember the feelings of pleasure and joy and exhilaration when you were a child. Those are the times when we knew what gives us our greatest joy of all. What did you love to do when you were little?

As we grow older we often lose the memory of what it is that gives us the greatest pleasure and joy of all. Commitments creep in.

- *The need to make money,*
- *The need to take care of children,*
- *Responsibilities of all types and characters take front seat.*

As it is the case with most people, many of us wind up doing and living our lives in ways that squash what it is that we truly wanted when we were a child.

- *How were you playful when you were little?*
- *What is the time when you remember laughing so long and hard it hurt all the way to the next day?*
- *What is it that you thought about when you were little, day after day?*
- *Maybe you did it maybe you didn't, but what thoughts do you remember having day in and day out?*

When we are young we remember. As we get older we forget. Allow yourself to become younger now year by year going back in time

- *Five years*
- *Ten years*
- *Fifteen years*
- *Marching backward in time year-by-year,*
- *Seeing the images*

- *Hearing the sounds*
- *Smelling the smells*
- *Feeling the sensations of each and every moment you experienced through those years*
- *March back in time,*
    - *year by*
    - *year by*
    - *year by*
    - *year*

until you arrive to the time when you are eight years old. Where are you now? Are you having a good time? Are you enjoying yourself? If not shift ahead in time until you come to a time, until you come to a memory that is truly and genuinely joyful to you as that precious eight-year-old, a time of supreme joy and exhilaration, a time that every cell in your body will remember.

- *What are you doing now?*
- *How are you feeling now?*
- *What thoughts are you having now?*

As that eight-year-old, now imagine that magically and mysteriously a genie appears to you. This genie who is quite beautiful in her own right stares you in the eyes with glittering sparkles exuding from her head, saying to you.

"My dear one, what is your wish today?"

Answer her. This powerful genie will grant your wish. It is a guarantee, but she does have one instruction for you, given that you are only eight years old. The instruction is this: You only have one wish.

- *Choose it.*
- *Stick with it.*
- *Stay with it.*
- *Want only this wish for now.*

Your genie tells you now,

"I promise you that in my own time as your genie I will grant this wish for you guaranteed."

- *If you waver*
- *If you waffle*
- *If you are wishy-washy my dear little one,*

Your wish will not be fulfilled. Make a choice. What is your fondest wish today my dear one?

Be specific. Detail out this wish if you want to make it so. If it is a wishy-washy wish, then that is what you will manifest - a wishy-washy outcome.

- ***Get focused.***
- ***Be disciplined.***

Though you are only eight, you can make it so.

Detail out this wish. What does it look like in material form? How will it feel when you get this wish fulfilled?

- *How will your body feel?*
- *How will your heart feel about being granted this precious wish?*
- *Feel it.*
- *See it.*
- *Hear it.*
- *Watch your wish unfold now.*

Worry not how long it will take. You are only eight and you have many years left in this life. There is plenty of time to make it so. Get your act together. Clarify what it is that you really and truly and genuinely wish.

You are only eight. You may be limiting this wish in some way or fashion Why limit it? Make the wish as large as you desire.

If you visualize a tiny, small, inconsequential wish it will be manifested in time. If you visualize a large and ambitious wish, it too will be manifested in time. It is not necessarily the case that the large wish will take longer than the tiny wish, for you see my dear child at eight years old, it is only a function of -

- *your focus*
- *your dedication*
- *your discipline*

to honor this most fond wish that comes from the depth of your being and your soul.

We knew what we wanted when we were a child. Remember that now. Honor how your life lived as an eight-year-old and invite yourself to march ahead in time now, just as quickly as you marched back in time -

year after

year after

year

- *growing taller and*
- *stronger and*
- *smarter and*
- *more courageous*

Allowing yourself to feel all of the feelings that permeated every cell of your body each and every moment through all of each of those years that have passed since eight years old to 9 to 11 to 15 - marching up

year after

year after

year after

year after

year after

year

Remembering the fondest wish

year after

year after

year after

year

Remembering the fondest wish that you knew so well when you were little, marching forward year after year until you find yourself in present time. Remembering now what it is that you always wanted and wished for when you were little.

Your work this week is to detail that wish out now for yourself. Detail it. Whatever that wish is - detail it out. Materialize it in your mind's eye

Focus on that wish now. From henceforth throughout

- *This day and*
- *Tomorrow and*
- *The next day*
- *For seven whole and complete days*

Hold this wish dear to your heart for it very well may be what you have forgotten as an adult.

Honoring this wish is a key step to recovering from the symptoms of Parkinson's. Many blessings and may your focus this week

Be disciplined.

Be determined.

Be declarative.

Best wishes

What Is Your Plan for Today?

Do you have a plan of what you want to be sure and accomplish today? Do you make a list of things to do and write them down on a piece of paper, making sure that you check off each item as you finish it?
Or, perhaps you have a list in your mind that you carry about of things to do today.

How long is your list? Does it have:

- Three items or
- Five items or
- Seven items or
- Twenty things to do today?

Is it typically the case that when you make your list everything gets checked off by the end of the day? If you are, give yourself a star for you are very unusual to be sure.

Five Steps to Recovery

What item on your list today is the most important item for you to accomplish? Look at the items on your list. Scan the list in your mind's eye if it's in your memory or go ahead and look at the item if it is on a piece of paper.

What calls out to you as being the one to do today that is the most important thing to do? Which one?

Are you having trouble deciding? That's not unusual.

- Decide.
- Decide now.
- What is important on the list?
- Make a decision.
- Make an executive decision.

Choose one item on your list to do. Focus on that item today until it is done.

Be decisive.
Be disciplined.
Make it so

The Place of "No Power"

Remember a time over the past several days when you were wishy-washy, a time when you just couldn't decide what to do next.
It was a choice of one thing or another or still another and you found yourself pontificating ...

minute after

minute after

minute...

Almost standing still, spinning about, and accomplishing nothing because you could not decide what to do next.

- Lost?
- Where?

In the land of nowhere that is where.

Remember that time -- we all have them -- over the last several days so that it's a recent memory. Connect with the feeling in your body of what it feels like to be wishy-washy.

- *Remember where you were when this happened.*
- *Remember the circumstance.*
- *Remember the challenge.*
- *Think about the people who were present if any.*
- *The confusion*
- *The sounds*

- *The challenge*
- *The frustration*
    - *Wanted to do this ...*
    - *Couldn't do that ...*
    - *Wanted to do this ...*
    - *Couldn't do that ...*
    - *wishy-washy*
    - *wishy-washy*

Connect with the feeling now.

That feeling is the feeling of low frequency. It is the feeling of "no power." It is the feeling that will get you nowhere because all of your energy has been dissipated into nothingness and inaction. Acknowledge the feeling for it is familiar to all of us –

- *the feeling of no energy*
- *the feeling of lethargy,*
- *the feeling of*
- *I can't decide why don't I just take a nap*

Remember the feeling. This is not the place to manifest recovery.
It is the place to manifest sleep

The Place of Power

Remember now a time over the past several weeks when you were sure about what you were doing, a place of clear thinking, a place when you made a decision about what needed to be done and you executed that decision without hesitation.

Think back. This may have happened in the last hour or yesterday or last week or perhaps even a few weeks ago.

Remember that time when everyone around you was confused and indecisive and wishy-washy. Remember a time when no one else could make a decision but you. You stepped forward. You made the decision for yourself and for others. You took action.

- No hesitation.
- No wishy-washiness.
- No waffling.
- No babble thinking about positives and negatives about this choice or that choice or lists about what's best to do or not to do.

Simply a decision.

- A firm
    - A clear
        - An unequivocal

Decision to do something to do anything - a big thing or little thing - it doesn't really matter. What matters is the feeling of decisiveness.

What does that feel like in your body?

1. Feel the energy surge.
2. Feel the power.
3. Feel the life force.

This you see is the place where recovery happens.

A Fond Memory from When You Were Eight Years Old

Slip yourself into a memory when you were eight years old, full of excitement and exhilaration, a time when you were in your essence, doing what you truly enjoyed doing as a child.

- *Eight years old.*
- *Find a time.*
- *Find a day.*

Remember what you were doing that day and that time when you were having the time of your life. The bucket was filling over and spilling with joy for it was full enough to contain all of the laughter and giddiness and care free

"I love my life"

feeling that you had that day in that time.

- *What were you doing?*
- *What are you doing?*
- *What's the activity?*
- *What are you wearing?*
- *Who else is there if anyone what are you touching, if anything?*
    - Is this a thing?
    - Is this a thought?
    - This is a place of joy.

Connect with everything associated with this place of joy.

- *The sounds*
- *The smells*
- *The tastes*

Feel this joy now. Allow it to soak into every cell of your body. Remember it. Re-kindle this place.

Doing what you are doing on this day and in this time may be a key, a clue to the mystery of what would be a good choice of about your fondest wish for today.

Focus Meditation

The following will constitute instructions for a 15 minute meditation I invite you to do each day for seven days in a row. You will not do the meditation now. Rather, I invite you to do the mediation at a time that you set for yourself.

The instructions are as follows: Locate a timer that can be set for 15 minutes. The timer will ring a buzzer when the 15 minutes are up. The timer could be a cooking timer or a cell phone or an alarm clock. Any timer will do, but it is important that you select a timer that will signal when the 15 minutes are up, else you will find yourself looking at the clock every 30 seconds. This defeats the purpose of the mediation.

Once you have your timer in place, find 15 minutes in the day when you will not be interrupted by anyone or anything. Turn off the cell phone (unless it is your timer). Unplug the phone. Be by yourself if you can. It works best that way.

Now, with your timer set, take one long breath in and out easily and effortlessly, not forcing breathing in and out but just allowing your body to relax and doing that consciously seven times more,

Allowing your body to relax
Sitting (or lying) in a place that's comfortable for you
Paying attention to your breath, in and out

Now that you are completely relaxed, lift up your non-dominant hand. If you typically write with your right hand lift up your left hand. If you typically write with your left hand pick up your right hand.
Now, look at your thumb. Take the 15 minutes that are now running in your meditation to focus your thoughts only on your thumb and nothing else.

When you notice that other thoughts creep in like.

- *"This is pretty silly"*
- *"I'm wasting my time"*
- *"When is my 15 minutes going to be up?"*

These are just the typical distracting thoughts that we always encounter when we attempt to focus on one thing. This is simply an exercise to discipline your thinking. That's all it is.

This meditation is a way of disciplining your mind to focus on one wish, one want,

- Not two
    - Not four
        - Not six
            - Not eight

One thought about the thumb of your non-dominant hand. Focus for 15 minutes.

Do it each day for seven days. Be surprised with each subsequent day how much easier it is for you to focus your thoughts on just that one thought.

It is a matter of discipline. You can do it. It will make a difference.

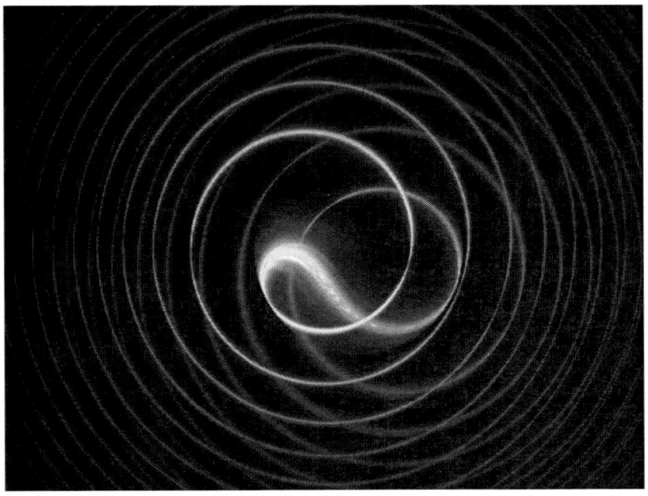

Step Two: Detail It Out

The focus today is on the second essential step that necessary for people with the symptoms of Parkinsons to recover. First however, I want to ask, how is concentrated focus on a single object working for you?

- *Is it difficult?*
- *Is it easy?*
- *Are you doing the meditation every day?*
- *Do you "forget" to do it?*
- *Do you avoid doing it?*

Speaking personally, it is a true challenge for me to focus on an object like a book for a 15 minute block of time without hearing thoughts creep in. Negative thoughts can easily dominate my thinking. Do any of the following thoughts sound familiar to you?

- *"I am old, you know."*
- *"I can never do that again."*
- *"Well, at least I'm not worse today."*

- *"It could be worse you know."*
- *"I am just destined to lead a life of misery."*
- *"I'm going to never recover."*
- *"Look what's happened to me."*
- *"What they say is not true. Recovery is not possible."*

There is a litany of thought forms that are connected to who we are and where we came from. We all hold negative thought forms in the tissues of our bodies.

Thought forms are not simply housed in our brains. They are housed throughout each and every tissue in our body.

Were you aware that each of us actually carries physical cells in our bodies from our mothers? I am not talking about an energetic (or mind reading) transfer of thoughts here. I am talking about the physical presence of a set of cells that actually came from our mothers. Scientists have proven that some of our cells are actually the physical cells of our mothers.

My mother certainly held negative thoughts and she expressed them frequently.

1. *I house some of her physical cells.*
2. *Those cells hold thought the negative forms of my mother.*
3. *The negative thoughts that she carried have been transferred to me.*

It is not that easy to be able to challenge and override the negative thought forms that we carry within each and every cell of our bodies. The challenge of negative thought forms does not end with the presence of a few cells from our mothers.

Negative thought forms are not just housed in our own cellular tissues. They are also carried by all of the bacteria and pathogens that circulate throughout our bodies that originally came from places outside of our body.

Many of you may be thinking,

> *"I don't have any bacteria right now. I had a few infections back years ago but the antibiotics did or good job of killing all of the bacteria."*

I can assure you that if you had an infection in the past - as almost everyone has had – bacteria are still circulating in your body. It is a rather distasteful topic for me to talk about, but I have seen pathogens removed from my own body. After all, we all have orifices that are exposed to the world.

Tiny microscopic pathogens are crawling all about the world. They creep into our bodies in one way or another through the food we ingest or simply from lying on the beach on a warm summer day.

There is no way around it. Pathogens like hanging around human bodies. They sometimes coexist inside us for years and years, huddled together, having a grand old time. Pathogens are living entities. They also hold thought forms and, as you might guess, these are not positive thought forms. They live in our bodies, feed off of our life force and influence our own thoughts. What a package!

Pathogens and bacteria both emit low frequency vibrations. When we are bombarded with negative thought forms for a sustained period of time this is one of the reasons why we will inevitably find ourselves beginning to feel worse and worse. We will ultimately manifest disease.

The bottom line is negative thought forms are really a way life for many of us for reasons outside our control. There is really no way around it.

I want to now move to address a question that pertains to this particular topic of negative thought forms that are associated with the medical diagnoses of Parkinson's. Here is a question I have received many times:

> *"I would love to believe that our bodies have the ability to heal themselves but how could this possibly be true?*

I do personally believe that all the wisdom about natural healing was in the minds and the hearts and the souls of our ancestors, our relatives who lived thousands of years ago probably in Mesopotamia and in Egypt. Our ancestors actually had the answers, but all that wisdom has been lost. Here we are, thousands of years later, trying to reconstruct the wisdom that once was common knowledge.

How do I know that people recover? The evidence shows that it's true. I have heard report after report from people who have the symptoms of Parkinson who are getting good relief from their symptoms. They are recovering. Some people have fully recovered.

They are doing it by listening to their bodies, listening to the signals their bodies are giving them and then responding accordingly. The people who are on the path to recovery are not locked into thinking they had this "illness" called Parkinson's that is a "disease." They do not rely on others to figure out what it is that is necessary for them to heal.

Rather, these pioneers are simply listening to the signals that their bodies are sending to them through the symptoms. They ask

> **What is my body trying to tell me?**
> **What does it need that I'm not getting it?**

The answer in the moment could be

- *Food*
- *Supplements*
- *Stress relief*
- *Trauma relief*
- *Toxin release*
- *Love*
- *Forgiveness*

There could be a number of different answers to that question. A combination of interventions may be required.

To be sure, many persons do listen to their bodies. They do experiment with one therapy or another. Yet, they may not be getting better, or at least they are not recognizing they are getting better. What's up Doc?

It is possible that patience is needed. It takes time for the physical body to heal. Still, what about the person who has explored lots of different alternatives and yet their symptoms still aren't getting better?

- *The root of the challenge does not likely lie in the supplements that have been taken or diets that have been chosen.*
- *It is not tied to the therapies that have been tried.*
- *It certainly doesn't it doesn't reside in the intention to get better.*

I believe that the challenge that most people confront resides in the wishy-washy nature of their thought forms. There is first of all an impressive body of hope for people about effective alternatives of therapies that can be pursued to get better. For success, there is a positive thought form that underpins recovery:

Recovery will happen for me.

This thought, the thought that it will happen, has to be sustained for a long period of time with clarity and succinctness. If the original thought of success is coupled with negative thoughts, qualifications or hesitations, the primary thought form becomes chaotic and messy. The thought

Recovery will happen for me

Does not have sufficient energy and power to manifest heal and wellness. It is vibrating at a frequency that is too low in order for the thought to be materialized in physical form.

When we focus our thoughts on the negative side of things, that is what we will manifest. Many people falsely think that when they say

> *"I want to eliminate my tremor"*
> *"I want to no longer be depressed"*
> *"I want my freezing to stop"*

They are placing energy on what they do not want to see happen. The thought forms focus attention and energy on the tremor, depression and freezing. That is precisely what is becomes materialized:

> *More violent tremors*
> *Deeper depression*
> *Stiffer rigidity*

When a person sets their intention to eliminate a tremor they are actually reinforcing the importance of the tremor in their life.
That thought form is a negative platform. It is not a positive platform. If I say to myself every day

This rigidity has to go

What word is remembered from this thought?

- *It is not the word "this."*

- *It is not the word "has"*
- *It is not the word "to"*
- *It is not the word "go"*

What your mind remembers
What you body remembers
What our consciousness remembers
is the word
"rigidity."

That's what gets reinforced.
That is what becomes manifested.
That is the thought form that has the energy.

It is not what we intended obviously, but we wind up manifesting precisely what we do not want to manifest.

How Thoughts Become Messy

Thought forms become messy and chaotic when we hold two diametrically opposed thoughts. I do this. Everyone does this.

On the one hand we say yes we want "A" to happen

1. *We want to sleep like a baby*
2. *We want to move like a ballet dancer*
3. *We want to walk like an acrobat.*

On the other hand and at the same time, we say we want "B."

1. *We hate sleepless nights and want them terminated.*
2. *We despise it when we freeze while walking across the street.*
3. *We are sick and tired of balance problems and want to stop falling.*

Allow this collection of messy thoughts to rattle around your head day in and day out. You will not be able to manifest much of anything other than messy, chaotic outcomes. When our thoughts about recovery are splintered, we will not manifest our true intention to recover.

Negative Thought Forms Dominate the Airwaves

I have had conversations and e mail with countless individuals who say the same thing. Here is the gist of what I hear:

> *When I search on the internet for answers, I get depressed. When I go to support groups, I get depressed. Everything out there is so depressing to me. It is so easy for me to lose hope, to be pessimistic about my chances of recovery."*

When negative thoughts dominate the airwaves – the television, the radio, the newspaper, the medication consultations, the health insurance claims processor, the internet, publications about Parkinsons - it is hard not to be sucked in by it all.

It doesn't even matter whether we hear it or read it. When there is a sewer of negativity floating all about us, it is a formidable challenge not to be sucked in and overwhelmed by the negativity of the sewer. We wind up floating down the sewer with everyone else. However horrible the feeling, it begins to feel normal.

I do not want to promote negative thought forms myself, but to make them conscious, permit me to summarize the words I found on a Parkinsons website which is typical of what you find on most websites about Parkinson's.

- *Parkinsons is a chronic, progressive disease.*

  The body is working perfectly. The symptoms give Valuable information about what the body needs to heal.

- *There is not much hope and, sadly, you will get worse over time.*

  People get better.

- The "illness" can become a source of much anxiety.

  The symptoms are a source of unbelievable creativity.

Feel into the depression that is associated with the untrue thought Parkinsons is a chronic, progressive disease.

Feel the optimism when you acknowledge that your body is working perfectly.

Feel into the lack of hope if you think you will get worse.

Feel the optimism when you acknowledge the evidence that people do recover.

Feel into the fear when you become anxious.

Feel the surge of energy when your creativity begins to re-emerge.

Feel into the helplessness of negative thought forms.

Feel the power of positive thought forms.

When you read the word "progression" do you immediately define the term to mean "I will get worse"? The term progression has a root which is "progress." Progress is an upward movement is it not?

I say yes, there can be progression with Parkinsons', but in a positive sense. That is to say, symptoms get better.

When I review the published research on Parkinson's, I find that even among people who are taking medicines of one type or another, 10% of the subjects do not get worse. I repeat, their symptoms do not get worse.

So, even in the studies that are well-funded with very large sample sizes what we see is evidence that all of the subjects do not get worse. It is simply not true. Some people taking medications do not get worse.

Research evidence also shows that at least one fourth of all persons who are diagnosed with Parkinson's are found not to have the "disease." This has been confirmed by autopsies. If one fourth of the persons who have the diagnosis of Parkinson's were misdiagnosed, why would they get worse? Of course, the answer is that they will get worse if they think they will get worse.

The Negative Belief Template of Parkinsons Disease

I think it's a mistake for anyone to trap themselves into the label of "Parkinson's Disease." It imposes a negative belief system that is connected to an army of negative thought forms that can stick to you like super glue.

What army negative thought forms am I talking about you ask? What follows is the collection of words I pulled from a three paragraph description of Parkinsons Disease.

- *Disability*
- *Chronic*
- *Progressive*
- *Illness*
- *Anxiety*
- *Not possible*
- *Level of disability*
- *Disorder*
- *Side effects*
- *Main concerns*
- *Hallucinations*
- *A life of adjusting*

That is a mouthful for just three paragraphs. Let me now list the positive thoughts, the words of hope and faith.

-

No, I am not writing with invisible ink. The number of positive thought forms numbers zero. There are no words of hope on the page.

I work every day at Parkinsons Recovery to override the negative belief template that dominates the air waves now. It is a formidable challenge. It is particularly difficult for anybody who has been diagnosed with the symptoms of Parkinson's because that belief system that Parkinson's is chronic and progressive is powerful.

What this particular description of Parkinsons does is to offer the conventional explanation that has been hanging around for decades. These thought forms are standard. You will find them on webpage after webpage out there in the Internet ethers.

The negative words and thought forms simply reinforces the existence of all of the symptoms that are associated with Parkinson's. Clusters of negative thought forms have low-frequency. They will ultimately bring anyone down to a state of disease and illness.

If we focus our thoughts on what is called "Parkinson's Disease" that nests us into a negative belief template that is genuinely depressing. The depression itself is enough to insure you will get worse.

The negative belief template is universal. It is powerful. It reinforces negative thoughts that are unhealthy.

Let me be perfectly clear. Even thought you are holding the positive thought frame of recovery, you may also be juggling negative thoughts in your mind. This dilutes the focus on the positive intention. The positive intention to recovery never cumulates sufficient power and energy to materialize. The intention is diluted in form and substance. Nothing can happen in this diluted place of chaotic mish mash.

The Medical Model versus the Model of Recovery

One of the reasons the belief template for Parkinsons is negative is that the thinking follows the medical model of diagnosis and treatment. The western approach is to ask: what is wrong? Once the "problem" is identified, it is treated. Thus, the thought forms that guide the investigation of a problem in western medicine are by definition negative.

Medical doctors come under close scrutiny by medical boards and associations. They cannot prescribe any medication before a diagnosis is made. They must follow a carefully prescribed protocol.

1. First, they must investigate the source of the problem with medical tests (which can be very expensive).

2. Second, they must attach a diagnosis to the problem.

3. Third, they must prescribe a course of treatment.

Medical doctors cannot reverse the order of investigation, diagnosis and treatment. They are severely constrained by legalities.

Many people tell me that they are frustrated because their medical doctor does not tell them about alternative treatments and therapies that might help. There is a good reason for their reluctance. Doctors could be sued if they recommend a therapy that has not been shown by research evidence to be effective.

Fortunately, I am not bound by such legalities. As a researcher, I am trained to identify any modalities that people say helps. I cannot diagnose. I cannot treat. I am certainly not qualified to do either. My job is to gather information and spread the good news of hope.

This means I operate in a positive belief template. I can expose positive thought forms for myself and others.

Celebrate and acknowledge the training and skills of medical doctors. Let them hold the negative belief template for you. They are familiar with this system and its rules. You do not have to take on the negative thought forms yourself. You do not have to adopt the negative belief template. That is what medical doctors are trained to do. So I say, let them do their job.

- **There is no need to adopt the negative belief template. Let others do that for you.**
- **You can adopt and sustain a very different energy supported by positive thought forms**
- **Holding a sustained focus on positive thoughts will result in incredible outcomes.**

This promise comes with a guarantee from the universe.

If I were Only Back to "Normal"

Have you ever thought to yourself:

> *"I wish I were like normal people"*
> *"Other people don't have the problems that I have"*
> *"Poor me."*

It would certainly be unusual if you did not have these thoughts. Of course these are negative thoughts. But they also misrepresent the truth of what we as humans confront in our lives. .

Is anyone with the symptoms of Parkinsons really any different than other people? Perhaps you are thinking –

> *"The answer is obvious. Other people are not sick like I am."*

Here is what happens when you run with this line of thinking:

> *"Look of all the other people in the universe that can function "normally." They have wonderful lives. I don't.*

- *I used to play the piano but I can't play the piano anymore*
- *I used to ski. I can't ski anymore.*
- *I used to run. I don't run anymore".*

The truth about living in a human body is that everybody – I repeat everybody - comes into this life with issues of one type or another. We may see other people as

1. *Handsome*
2. *Beautiful*
3. *Successful*
4. *Happy*
5. *Healthy.*

It looks like it anyway. But what do you really suppose is underneath the appearance? A writer named Shakespeare over four hundred years ago made his contribution to literature by hammering one idea home:

> *Appearances never match up the reality of things as they truly are.*

It was true then and it is true today.

- *The people who are handsome and beautiful often feel ugly and unlovable inside.*

- *The people who are seemingly smart and successful often feel a deep sense of defectiveness deep inside.*

- *The people who are seemingly happy are often putting on a mask to the world. Deep inside they are depressed.*

- *The people who appear to be healthy may actually be confronting very serious health challenges that they do not want anyone to know about or may not even be aware of themselves.*

If you touch deep inside anyone, if you really get to know them, you will discover that they too are confronting some very serious and formidable challenges in their life. Or, if they are not confronting them, they should be! No one in a body escapes.

If you look at me, if you look at Deborah, if you look at our children we are handsome people. Our children are handsome people. They are smart. We are healthy. We live in a wonderful place on the Puget Sound. We love each other. But all things said, I can assure you that each of us has very specific issues that were dealing with that are very deep and very profound. No one escapes.

It just so happens to be that the symptoms of Parkinson's are the issues that you have been dealt with. Everyone on the planet has issues. It just happens to be the case that the symptoms of Parkinson's are the issues that you have been dealt.

I suggest that you say to yourself

> *"Okay. I'm like everybody else. It just so happens that what's happening to me right now is what I need to deal with. This seems to be what I need to work through right now. Others get dealt a different stack of cards."*
>
> *"Thank you very much. Now I am going to get to work figuring out what this is really all about for me."*

Focus on the End Result

Take positive action that is supported by positive thoughts. Focus your thinking. Focus on a single intention without wobbling back and forth, letting negative thought slip in and muck things up. With focus, the intention marshals up energy and power.

Then materialize the intention. See clearly what you need to do. Materialize your intention. Detail it out.

Manifestation of your intention happens much quicker when you have materialized it. The more detailed the intention, the quicker it will be manifested.

Five Steps to Recovery

How do you maintain clarity for your intention? Focus on the desired outcome, on the goal. What is the end result you want to achieve?

You do not want to focus your thinking on the steps that are necessary that lead up to the accomplishment of the goal. Be clear at the outset about where you want to go. Once the outcome is clear, you can figure out how to get there.

Think about it. What happens when you decide you want to make a trip somewhere, for pleasure or business? Don't you pinpoint the destination first and then look at a map or at airline schedules to figure out how you're supposed to get there?

It works the very same way with our thought forms. The most powerful force in the universe is thought.

- *When thoughts are crystal clear*
- *When thoughts have clarity*
- *When thoughts express clear intention the intensity*
- *When we hold those clear thoughts for a prolonged period of time*

the universe will deliver by manifesting your thoughts in material form. If we qualify our thoughts with doubts and hesitations, they will not materialize because they had been diluted with wishy-washy thinking.

To summarize, two keys to recovery are to

1. *Focus and concentrate your thoughts*
2. *See it done. See the goal accomplished*

- **the longer you can see at**
- **the stronger you can feel it**
- **the louder you can hear it**
- **the faster your heart's desire will materialize.**

The problem may not be that you do not hold clear, positive thoughts. The problem may be that they lack the energy to materialize because you are not holding them for a long enough period of time. Thoughts that flip in and out of our minds do not have sufficient intensity. Concentration of focus is lacking.

Remember, we can only think of one thought at a time. Make it positive. The trick of it is to work on;

One intention
One goal

Keep that the focus of your thoughts. That is it. That is all you do. Focus on a single intention. And this you see is the answer to the biggest mystery of the universe.

Clarify Overview

This week's meditation will help you clarify your intention for recovery. What is your intention? As you know it is paramount to select

One intention
One wish
One want

Whatever you might call it, select one. Which one will it be?

- *Perhaps you have been thinking that the one intention you'd like to focus on is to eliminate rigidity.*
- *Perhaps you may be thinking that the one intention you would like to set for yourself is to stop the tremoring.*
- *Still others may be thinking the one intention is to terminate the depression.*

There is a fundamental flaw with these thoughts. They give energy to what you do **not** want to see happen. Of course you are saying that you do **not** want them, but the universe does not acknowledge the word "**not**." For some reason, the word "**not**" is not found in the dictionary of the universe.

The only words that are recognized, encoded and acknowledged are "tremor," "rigidity" and "depression." As it turns out this is what is manifested in material form. As the universe listens, this is what the universe hears.

- *I wish not to have rigidity.*

What does the universe hear? Only the word **rigidity.**

- *I wish not to be depressed anymore.*

What does the universe hear? The universe only hears the word **depression.**

That is what the universe believes you want and that is what you will get.

- *If you wish **not** to have tremors, you will be guaranteed to have tremors.*
- *If you wish **not** to be rigid anymore, you will be guaranteed to be rigid for the rest of your life.*
- *If you wish **not** to be depressed, it is now guaranteed you will be a depressed cookie for a long, long time.*

You see, intentions that are couched in negative terms simply give those intentions power. They give them energy. They give them force. They focus attention on what it is that you do not want.

Each word taken by itself has a very low frequency. But because the words are most likely repeated frequently, they attain enormous force and energy which ultimately is realized and sustained in material form.

Think about how you feel when you say to yourself

- *I want to stop feeling bad.*
- *I want to be able to at least walk two steps instead of just one.*

Those types of statements - those limiting statements - have very low energy attached to them. In just saying them now, I feel a plunge in my own energy. I may not have enough humph now to finish this meditation.

To summarize, when we talk about what we don't want to see happen in our life, that's exactly what does happen. When we limit ourselves to two steps instead of limitless steps, that is what turns out being manifested in our lives.

How do thoughts work for us rather than against us? With the "right" thinking we will manifest what it is that we wish to manifest in our lives. Yes, it is possible to recover with the right thoughts. But our does it work?

So you ask,

okay
okay
okay

How do I think then about what it is that I wish to see happen for myself and perhaps for others as well?

I have a meditation task I will now ask you to ponder now. First however I invite you to place yourself in a comfortable position, whatever that looks like to you.

There is no right or wrong position here. When meditators talk about meditation, you oftentimes see them in lotus form with their legs crossed and back straight. Well, that may work for some people but for many it's a big flop. It's not necessary to be in any particular position in order for a meditation to succeed for you.

The right position for you is the position that helps you feel comfortable and relaxed. Perhaps that's lying down on a couch or on the floor wherever that might be -

- *be it a hammock*
- *be it floating in a swimming pool*
- *be it sitting in a car and relaxing.*

Wherever that place might be, allow yourself to simply go to there now, that place of relaxation.

Five Steps to Recovery

Take a long deep breath in and out. As you breathe, I invite you to do it mindfully, not something you have to do. This is something you want to do for yourself,

a way of slowing down time,
a way of the re-acknowledging the magic and mystery of life itself.

With each breath we witness a miracle. Breathe in however deeply it feels comfortable. Pay particular attention to the air as it leaves your body, allowing all of those difficult feelings to simply dissipate into thin air.

With each breath allow all the uncomfortable feelings you may have been having today to dissolve. Do not count the breaths. Simply pay attention to the air, the breath of life as it moves inside your body, touching each and every cell throughout your body. Allow the air to exit ever so effortlessly.

As you breath notice any muscles that twinge or are tight. With your intention, invite them to loosen up.

- What's about to happen will be enjoyable.
- It will be life giving.
- There is nothing scary about what's to come.

Make a quick scan from your toes all the way up to the tippy tip top of your head, checking to see if there's tenseness anywhere

- *Behind your eyeballs*
- *Inside the innermost part of your throat*
- *Chest*
- *Trunk*
- *Buttocks*
- *Thighs*
- *Knees*
- *Hamstrings*
- *Ankles*
- *Little toe*

Scanning

up and down
up and down

your body everywhere, doing a quick check. No big deal here. Just check to see where there might be tension that no longer needs to be present. Acknowledge the magic and the mystery of this home of yours we all call a body.

I now invite you to muster up in your imagination an activity that you have given up either recently or a while ago because of the symptoms of your Parkinson's. The activity that comes to mind could be a small thing like

- *Buttoning a button or*
- *Brushing a difficult to get to tooth or*
- *Combing the back side of your head or*
- *Going to a party and dancing or*

- *Skiing on the slopes of Mount Hood in Washington or*
- *Bicycling or*
- *Marathon running or*
- *Playing the piano or*
- *Playing the guitar or*
- *Playing the flute or*
- *Playing whatever instrument you once played when you were younger.*

Muster up that activity in your imagination now, an activity that you fondly remember and you desperately miss.

- *What activity gives you immense pleasure that you have given up?*
- *What activity do you find yourself thinking about during the day?*

Oh, I wish I could do that again.

- *What activity do you sometimes dream about at night?*

As you think about various activities that you have given up - and everyone has given up activities regardless of the symptoms involved - what activity gives you intense and passionate energy? What did you really love to do?

- *Sailboat racing?*
- *Car racing?*
- *Pool playing?*
- *Card playing?*

What was it that you really loved to do that for whatever reason you have decided to give up. The activity you're thinking about now - if it's giving you great joy inside - is the activity I invite you to now work with in this meditation of ours. Put it aside in your imagination for a moment.

I now have a second request. Create in your imagination a very, very large and elaborate television screen, as large as you would like because, after all, the TV screen is only in your imagination.

Place the screen in front of you. Place now in either your right or left hand a device you can use to turn the TV on and off and to switch channels. Make the controller however you wish it to look. Perhaps making it in gold would be your preference or silver or chocolate. It's your controller. Make it whatever shape and color you wish.

Turn the TV on now. The show that you are now watching is the activity that you have given up for whatever reason. This is the activity that you passionately would like now to reclaim.

See yourself doing that activity now just as you did that activity when you so enjoyed it. It is your movie. Start it now and allow the activity to unfold.

Where are you?
What are you wearing?
What are the sounds that you are hearing now?

Put whatever in this scene whatever and whoever needs to be present as you watch yourself doing this most joyous activity. Remember. It's your movie. You're the director. Embellish it however you so desire.

Add colors where you might wish to add colors. Decolorize it if that's more fun.

Who else is with you now?
Who are the other actors?

Listen to what is being said by you and by others if others are present. Listen to the thoughts.

- *Are you limiting yourself with this activity in any way? Why?*
- *Are you a little slower than you were when you truly enjoyed doing this activity? Why?*
- *Do you have worrisome feelings as you watch your television screen? Why?*

This is the activity that you truly, truly enjoy doing and relishing. This is your movie. See yourself doing what you truly enjoy doing, precisely the way you loved doing it. Give yourself the body that you want to have to do this activity again now.

- A body that's free-flowing
- A body that is fluid
- A body that is agile
- A body that is young and competent

- A body that possesses a surge of energy to accomplish whatever needs to participate in this activity effortlessly.

Nothing can stop you. You have all that you need to do this activity now.

Stay with it. As your TV program unfolds it may change character. Allow it to do so. Join the ride.

> *Engage the fun.*
> *Feel the joy.*
> *Stay focused on just this one activity.*

You may have been tempted to change the channel to another activity. That's understandable. We all do it. Switch the channel back to the channel you started with.

Stay on track. Stay focused. Watch yourself as you do what you truly love to do on your TV screen. Keep your fingers off the controller unless you need to turn up the volume or adjust the brightness.

- *Is your picture fuzzy?*
- *Is it a little unclear?*

There are dials on your controller that can adjust the focus so that it is precise, so that you can see the details of this activity. Your controller also has a focus.

As you're watching yourself do this activity, focus in on your screen on the top right section of the screen.

Focus in

Focus in only to the top right section of your screen. What is the detail there?

- *The shape of the objects*
- *The colors*
- *The textures*

If they're not on your screen now simply invite them to come in. Take in the richness of an experience you have so thoroughly enjoyed doing for so many years of your life.

Take it in. This is a moving picture, not a static picture. See yourself moving

Gracefully
Effortlessly
Joyfully

Receive now dopamine rushes from the endless laughter and joy that pervades each and every cell of your body.

This is the channel you have been looking for on your television set for decades and decades. You have at long last finally created it in your imagination.

- *Stay on this channel.*
- *Watch only this movie.*
- *Stay on track.*
- *Stay focused.*
- *Watch this activity and this television channel each and every day.*

You'll find that the details will become even more vivid with each additional day that you watch. The activity itself may transmute from one thing to another as was the case when you were doing the activity.

Our thoughts do ultimately result in physical reality. If we have fuzzy, unclear, opaque concepts of what it is that we would like to see happen, that is precisely what we manifest: something that is pretty vague, fuzzy and unclear.

The more details you can visualize about the activity you are now ready to reclaim, the quicker it will be manifested. The vividness in your imagination of this thought gives it power. It sets into motion all of the activities that are necessary for you to be able to do what is needed to reclaim the activity for yourself.

Hold this image day in and day out and the dream will ultimately be manifested in physical form ...

- *Because this thought has power*
- *Because you are not channel surfing*

- *Because you are holding firm to the control button on just this one channel*
- *Because you are adjusting the clarity and the focus of what it is that you'd like to see unfold*
- *Because it has become a vivid image to you*

It has become real in thought form and because of that, it will be made real in physical form.

If there's waffling, you have to add time. If there's too much wishy-washiness it may never happen. It's the intensity and the declaration of intent that will make this happen.

I invite you to watch the station every day for 10 to 15 minutes. Turn it on when ever your heart so desires. Detail out what it is that you would like to see happen in your life.

When that is materialized, then and only then I invite you to change channels to the next activity that you would like to reclaim.

May you find energy, joy and endless satisfaction in reclaiming your life force this week as you stay tuned to the channel of your heart's desire.

Detail Out Your Goal

Find a timer from your kitchen or your watch or your cell phone and simply set it to our 15 minutes. A timer is a good idea because you will not need to spend too much time on this meditation. It is a much better plan to do it every day for 30 days. Make it a daily ritual.

Do this mediation the same time every day if possible. Do it the first thing in the morning when you get up is a great idea, but any time of the day is good as long as you do it. If you do this meditation every day you will be amazed at what will unfold for you. Doing it every day speeds up the recovery process.

This meditation will not address what needs to happen in order for you to be able to accomplish the intention that you are setting. This meditation will not address specific decisions with regard to choosing Supplement A or supplement B seeing health practitioners C or health practitioner D.

This mediation will address detailing out and making clear the intention you wish to set for yourself. Once the intention becomes crystal clear, it will have sufficient energy to manifest. I guarantee you will find ways to make it happen one way or another.

You will experiment using

Different modalities
Different people
Different health care practitioners

You will figure out a way to manifest your intention. The specifics will work themselves out in time. To manifest the goal, you have to make the intention crystal clear in every detail.

- *Find a relaxed place*
- *Sit down if you are standing*
- *Get comfortable*
- *Slip into a relaxed space*
- *Place both feet on the floor*

Quiet the thoughts in your mind. Slow the rapidity of that rapid fire of thinking down. Take in seven, slow breaths effortlessly. This is doesn't have to be a big deal.

Take a breath into your lungs. Focus on its character. What does it feel like?

Exhale that air, counting at one and then allow yourself to count six more breaths. Whether they are short or long doesn't matter.

What matters is to focus only on the breath. As you do, allow all the tension in your body to dissolve little by little, noticing if there's any tension or tightness in:

- *your stomach or*
- *your shoulders or*
- *your toes or*
- *your thighs.*

Just let it go

- *No reason to be tight*
- *No reason to be anxious*
- *No reason to be upset*
- *No reason to be angry*

Let all of that go and relax your body. After you finished your seven breaths if still feels like you are tense, give yourself the treat of another seven breaths. There is no reason you must hold your initial meditation to just the first seven. Seven is the magical number of the universe. It is:

- *The number of days in a week*
- *The number of continents*

There is something special about focusing in on seven breaths. Do it in sets of seven until it feels like you are able to

Relax your body
Quiet your thoughts
Be present to yourself

Focus now on getting back an activity that you have given up because of your symptoms. Whatever comes to mind first is going to be the activity that would be best for you to meditate about today.

Make it a specific activity. The activity

- *Could be playing bridge with your friends*
- *Could be skiing*
- *Could be sail boating*
- *Could be playing the piano*
- *Could be playing the guitar*
- *Could be singing*
- *Could be visiting a friend*
- *Could be traveling*
- *Could be going to the drugstore*
- *Could be going to the grocery store*

It could be any number of activities, small or large, that you given up. It could be an activity someone else is performing for you now that you would like to reclaim for yourself. What activity have you given up that you want to reclaim for yourself? Allow that to float to the top.

Whatever you are thinking about, whatever is floating to the top is going to be the activity that will be truly and genuinely helpful for you to meditate on today.

The magic of the second step is to materialize this activity. See it in your mind's eye.

Using your imagination,

Five Steps to Recovery

1. *Place in front of you a television set with a blank screen. Make the screen as large as you would like since this television is in your mind's eye.*
2. *Place a TV control device into your hands.*
3. *Push the "on button"*

The TV program that you are about to watch is the program of you doing the activity that you have given up, the activity that you would like now to reclaim. Watch yourself now on the TV doing this activity. Make it ever so detailed.

- *What are you wearing?*
- *What shoes do you have on?*
- *What shirt or dress?*
- *What does your face look like?*

Watch yourself now moving into this activity

- *Who are you with?*
- *Who are the people that you're with, if anyone?*
- *Who is here?*
- *Where are you now?*
- *What does this place look like?*
- *What are the colors?*
- *What are the smells?*
- *What do you hear?*
- *How are you feeling?*
- *What is your body doing?*

You can make your body do whatever it needs to do to do this activity Make it so now. Move into the activity. Experience it in its fullest, now.

- *Hear it*
- *See it*
- *Touch it*
- *Smell it*
- *Taste it*
- *Feel it*

This is your movie
This is your life
Reclaim it now

You are focusing now on the positive, on what it is that you want to manifest in your life. Stay focused on this activity. Keep the TV tuned to the same channel. Carry through this activity to its fullest. Experience this activity in the way that you want to experience it, however that might be.

As you experience it, feel your body in a place of full balance, health and wellness, a body that is able to do this activity just the way you have always done it in your life,

- *Giving you pleasure*
- *Giving you joy*

The more concrete you can make this activity for yourself now the faster it will materialize.

- *Make it specific*
- *Detail out the place you are at*
- *Every detail counts*
- *Every detail counts*
- *Materialize this wish on your television set*
- *Allow this activity to unfold in its fullest with*

    - *No hesitation*
    - *No constraints*
    - *No limitations*

Do it the way you want to do it. Give your body the permission to be all that knows how to be now

Good work
Good focus
Good activity
Close that focused attention out
Give yourself a break
Take a breath
The work is done for the day

Put the meditation on your calendar to do tomorrow and the next day and the next day, the same activity. You may see yourself doing the activity in different places or in different ways. This meditation will help you materialize your positive intention.

In this space you are not focusing on the limitations or problems or symptoms. You are focusing on all that you will become. Allow this thought form to manifest to its fullest and most complete extent

- *Focus*
- *Detail it out*
- *It will happen*

Whatever you can do or dream you can do, begin it. Boldness has genius, power and magic in it. Begin it now.
Goethe

Why wait? In the waiting you will make the opposite of your dreams to happen.

The Energy of Negative Thoughts

I invite you now to do a quick survey of the thoughts you have had today until the moment you have started listening to this meditation.

Pretend you are listing each and every thought you have had since you woke up this morning on an imaginary computer. There will be a very long list, even if you have been up only for a few minutes.
Just imagine you are writing a theme paper that is

- *hundreds and*
- *hundreds and*
- *hundreds of pages long.*

Each and every thought is listed as it was stated in your mind.

Scan now these thoughts. Mark a check beside each thought that is connected to

- *not wanting to feel the way you are feeling,*
- *not wanting to have the discomfort the difficulty*
- *the challenge*
- *the pain*
- *the depression*
- *the difficulty with moving*
- *the difficulty being able to accomplish any small task*

Mark any thought on your list when you said to yourself

- *"I don't like this"*
- *"I don't want this"*
- *"Stop this"*
- *"This is terrible"*
- *"Is this the way it's going to be always?"*

Check all of those thoughts you have had today that are just like these negative thoughts.

1. I do not
2. I do not
3. I do not
4. I do not

Check all of those thoughts that you see on your list.

Guess what? Most people discover there are many, many checks beside thoughts that are negative on their list. In fact, the number of negative thoughts can consume over 50% of the thoughts that you've had today until this moment.

In so thinking this way – in negative terms - you are giving energy to maintaining the state that you find yourself in now. When you say

- *"I do not want ...*

the only thing that happens is that you reinforce and insure that condition will be maintained or even become worse.

Five Steps to Recovery

I invite you today and the rest of the week to notice each time in your thoughts you say:

- *I do not want*

In so saying that, you give that negative thought more energy which is the opposite of your intention!

You say you do not want to have movement problems? That in itself, that thought, is a negative thought that sends energy to that condition. That type of thinking will guarantee that the movement problem will be sustained and continued and perhaps even worsen.

I invite you to notice whenever you use the words

- *I do not*

In your thoughts or even as you vocalize your thoughts to others.

Pinch yourself. Remind yourself.

Wait a minute. This just gives all of this energy to what I do not want to see happen.

Eliminate

- *I do not*

from your vocabulary and you will see a positive shift in your symptoms.

Thinking about Trying out a New Therapy?

Think about a therapy you have recently decided to experiment with. Perhaps the therapy was to try

- an herb or
- an essential oil or to see a healthcare practitioner like
    - a naturopath or
    - a psychologist or
    - a neurologist.

Think back recently. Perhaps this happened for you just in the last few days or perhaps it's been several months. Just think back to the most recent new therapy you've decided to experiment with - whatever that might be - and whether or not it might have cost anything or not.

As you connect with and remember that choice, I invite you to remember the thoughts that were associated with making that choice. Did you have doubts about whether or not it would help? Did you say to yourself,

Well it can't hurt.

Did you say to yourself,

> *Well I have my doubts you know. The people who do this kind of work are a bit wild and screwy and weird.*

Or, do you remember yourself saying,

- This is going to help me.
- I know this is going to help me.
- There is no doubt in my heart or mind but that this is what I need now.
- Maybe it's not what I'll need in a year but by Jove, I'm going for it because I know this is what will help me now.

What was the thought that you had? Remember.

If you had doubts I will make the prediction that the outcome would not have been particularly positive. If you were sure it was a good decision I will make the prediction that just as expected, the result was wonderful.

There is in social science research something called a placebo effect which is a complicated way of saying that if you expect something will happen it will happen. And, if you expect it will not happen, it will not happen. Our thoughts determine the outcome.

Next time you select a therapy to be experimented with, be very attentive to the thoughts that you tag to that therapy. If you are hesitant, scratch it.

Choose a therapy you can be positive about and you will see a huge shift in how you feel.

The Value of Detailing What You Would Like to Manifest

When you detail out and are specific about precisely what it is that you would like to see happen, it gives that aspiration energy. That type of thinking leads to a materialization of that aspiration.

Think about a movement that you have that has been troubling you recently,

Perhaps it is walking across the street or

Perhaps picking up something from the floor or

Perhaps combing your hair or

Perhaps turning in bed or

Perhaps fluffing your pillow

Whatever that activity is – and there may be many - think and focus now on just one activity that is creating challenges for you.

With this activity in mind, detail out now precisely what you would like to see that activity

Look like and
Feel like and
Be like for you.

See it the way you would like for it to be. See yourself in all of your fluidity, just as you were when you were two years old with a rubbery flexible body that can do anything.

In your imagination see exactly what you would like this activity to be for you. Detail it out, every fraction of a second of this activity.

- *Feel it*
- *See it*
- *Hear it.*
- *Smell it.*
- *Eat it.*
- *Become this activity*
- *Colorize it.*

The more detail that you can give to an activity that you want to reclaim, the far more likely it is to be materialized sooner rather than later. You give that activity energy by detailing and clarifying what it is that you would like for it to be for you now.

Give it energy. Detail it out and that activity will shift for you because you have clarified it in your thought form.

What Does it Really Mean When You Have Difficulty Moving?

I have a question for you to ponder today. Since you began having the symptoms of Parkinson's have you been giving up certain activities that have been difficult for you to do?

Perhaps visiting friends
Perhaps doing yard work
Perhaps playing tennis
Perhaps sailing
Perhaps playing a musical instrument

When this happened, my question for you is did you say and did you think to yourself

Oh heavens, this is another thing I have had to give up.
I wonder what's next?

Are you keeping for yourself a mental list of things you've had to give up because you've been having challenges and difficulties?

If so, you are into a negative thought frame, into a belief template that is not true.

Somebody that does not have the diagnosis of Parkinson's who has difficulty moving today or even freezing would say,

"Oh, that's funny. I stumbled or I froze. Hum, that is strange.

They would laugh it off and think to themselves.

> *Oh - I better eat something or*
> *I better get a little more exercise or*
> *It's no big deal*
> *It will pass.*

In other words, there is no negative thought that's associated with that particular challenge.

If there is a negative thought that is given to that particular challenge, the mere existence of that thought alone gives more energy to the movement challenge you are having. It insures that there will be more difficulty with that movement in the future.

When you hear yourself with that type of negative thinking, simply pinch yourself and make a decision. The decision is to say,

- *Wait a minute*
- *Wait a minute*
- *It is not true I destined to get worse.*

> *I am ready to reclaim that activity after I get the therapies or treatments or I do the things for myself that I need in order to reverse the challenges that I'm having.*

That positive thought form will yield a shift in the challenge that you're having. Guaranteed

How We Obstruct the Recovery Process

We obstruct the recovery process when we immerse ourselves in the detail of the process of getting done what it is that our heart desires rather than focusing our thoughts on the outcome.

Think about it. When you decide that you want to take a trip or a vacation, you first decide typically where it is that you want to go. Once you make that decision, you then figure out the best way to get there.

It's the same with manifesting recovery. The key is to focus your thoughts on the end result, on the positive outcome that you have decided to materialize.

Detail that positive outcome out in your mind's eye and imagination every day, every moment. If you find yourself being inpatient because somebody has been too slow or is taking too long at the grocery store line or for whatever reason holding you up, take that as an opportunity and Invitation to detail out what it is that you have declared you wish to make happen.

Detail it out to the finest degree. Every element of what it is that you would like to accomplish needs to be seen in your imagination and felt.

The more you think about this outcome,
The more you detail it out for yourself,
The more you …

- *Feel it*
- *See it*
- *Hear it*
- *Taste it*
- *Sense it*

the more likely it is to happen. This has nothing to do with how you make it happen.

The process of it all will automatically unfold. The answers will be staring you in the face because you have set the outcome clearly and without equivocation and with declarative assertion.

Make an executive decision to detail out what it is that you want every moment you can and it will happen all the more quicker.

Step Three: Stay on Track

This chapter discusses the third of five steps that are critical to being able to launch a successful program of recovery that in the long run leads to significant relief of symptoms from Parkinsons.

You may be thinking that the third step is a specific list of therapies.

- *Perhaps I am going to talk about a herb or supplement that you must take to recover.*
- *Perhaps I am going to talk about some great concoction of antioxidants.*
- *Perhaps you are anticipating I will tell you that for $59.33, I'll send you this magical concoction of anti-oxidants. We have 1,921 testimonials of people who are feeling better because they took this supplement.*

That's not what I'm going to be discussing and that is not at the core of the recovery process. Of course, antioxidants may help, but the question here hinges on addressing the root factors that influence recovery.

What then is at the core of the recovery process? The power of thought unleashes the recovery process which is waiting to be activated by you. When thoughts are focused and detailed, they are anointed with power.

The third step of the program is very simply put to "Stay on Track." Think about what happens when you're driving on the highway. Perhaps you are in the habit of speeding up to really racy speeds of 80 to 85 miles per hour and then slowing down to speeds of 30 or 35 every minute or two. It's a familiar, frantic energy. You want to push through the traffic, but you suddenly get jammed up and have to slow down, almost to a standstill. The reality is this is a very inefficient and very costly way to travel.

Consider a second example. You are riding the railroads in Europe. Your goal is to travel from London, England on the train all the way to Nice in the southern part of France. The trip is a vacation so anything goes.

On whims, you start making side trips. Whatever whim might strike you fancy you jump the train here and there and spend a day or two or three or a week at one interesting city or another and, I might add, having a great time.

This is living life in its fullest. Life becomes a process where whatever seems to come up is what you need to do.

- *It is pleasurable.*
- *It is fun.*
- *It is engaging.*
- *It is energizing.*

But may I point out that while the side trips have been fun you haven't yet arrived at your destination. You are doing a lot of other great and interesting things but you haven't arrived yet. You haven't attained what it is that you set out to attain. You keep delaying arrival at your destination.

- *You have to sustain the same thought long enough for it to materialize.*
- *You have to keep it bubbling in your mind.*
- *You have to hold onto it for dear life.*
- *You have to make it your best friend.*
- *You have to house it in your mind*
- *You have to presence it moment to moment, day-to-day in order for the thought to materialize.*

When you veer off track, you will not be able to attain what it is that your heart desires. It is a lot like a cuckoo clock which have pendulums that swing

- ***back and forth***
- ***back* and *forth***

 Robert Rodgers, Ph.D.

Five Steps to Recovery

The clock does keep good time but the cuckoo clock never goes anywhere.

- It waffles
- It vacillate from side to side does
- It tells you what time it is

Time simply goes around and around and around every 12 hours on the clock. The pendulum swings back and forth but it doesn't really travel anywhere other than its own backyard. Of course that's what a cuckoo clock is designed to do.

When you have a clear intention about what it is that you'd like to be able to accomplish for yourself, how are you sabotaging yourself with negative thoughts? What is being sidetracked by your negative thoughts?

I want be clear about what wishy-washy thinking is really all about. Consider the activities you can do that underpin your recovery.

1. *You envision an activity for yourself that you want to manifest.*
2. *You do focus meditations every day.*
3. *You focus your thoughts.*
4. *You detail out more and more clearly exactly what it is that you wish to manifest*

It is not as easy as it sounds. As you have been doing mediations, have negative thoughts hit you from out of the blue? Is it as if there is nothing you can do to stop negative thoughts from infiltrating your mind? Perhaps some of these thought will sound familiar:

- *I'm going to get worse*
- *I deserve to get worse*
- *I tried everything - I mean everything - and nothing helps.*
- *I have to do everything all by myself*
- *There is no one to help me*
- *I'm all alone*
- *I don't have the money to pay for therapies*
- *The people who do recover have money*
- *Times are tough*
- *I can't do it*
- *Forget it*
- *My friend has been working on reversing her symptoms of Parkinson's now for years and she is still at it. She does everything – food, exercise, support groups – and she still shakes. If she can't do it, I certainly can't do it.*
- *I really don't think it's going to happen for me*
- *I can't*
- *I won't*

The creep, creep, creep of negative thoughts is haunting.

- *Yucky feeling after*
- *Yucky feeling after*
- *Yucky feeling after*
- *Yucky feeling*

- *Negative thought after*
- *Negative thought after*
- *Negative thought after*
- *Negative thought*

All of this wishy-washy muck discredits and dilutes the energy behind the positive intention that you've set. The rapid fire of negative thoughts set off a chain reaction. When the negative thought process is activated,

The energy is attached to the negative thought forms,

so that is what is manifested and materialized.

The energetic vibration of negative thoughts is low. When we embrace negative thought forms, we dive into a gloomy, dark place. It is depressing to live with a continuous battery of negative thought forms.

One negative thought leads to another until yes, you are on track, but not on the track to recovery. You are on the track to manifest everything that you do not want to see happen.

> Hesitation and wavering weaken your resolve. They get you nowhere.
>
> Hesitation and wavering dilute the intention you have set to recover.

- *What is needed to carry a clear commitment to recovery*
- *What is needed is a strong will.*
- *What is needed is decisiveness.*

Staying on track with positive thoughts is the difference that makes a difference. Positive thought forms emit high energetic frequencies. They possess the power to materialize your goal far more quickly.

When you do not stay on track, I can assure you that your intention will not materialize. It will not materialize because the thought does not have the energy to translate into physical form.

Make the power of positive thought forms work to your advantage.

- *It feels better.*
- *It is energizing.*
- *It is powerful.*

Blocks to Recovery

You Do Not Know What You Want

Why is it that people do not manifest positive progress toward their recovery if they have the symptoms of Parkinson's? One reason is they really don't know what they want. They are wishy-washy.

A person may say in the first hour of the morning

"Well, I'd like to be able to take a walk today."

Thirty minutes later –

"No, I don't feel that well. I think what I'd like to do is to go to the corner grocery store, buy some organic veggies and make a delicious organic casserole with fresh vegetables"

Thirty minutes later –

"I don't have enough time to go to the grocery store. Instead ..."

You get the point. This person really doesn't know what they need or want. Decisiveness and a strong will transform split intentions into a powerful energetic force.

Holding the False Belief that Recovery is not Possible

Second, you don't believe that recovery is possible. You really don't believe it, for understandable reasons I might add. There are certainly many, many people out there in the universe who are saying that it is impossible to get relief from the symptoms of Parkinson's. That thought form bleeds out everywhere –

- *In writing*
- *In talk*
- *In gossip*

If people believe they're going to get worse and feel worse. I will guarantee you that is what will happen. Guaranteed.

If people believe they will get better, they will get better. That is what will happen. Guaranteed.

- *It does not happen in a minute or day*
- *It does not happen instantly.*
- *It does not happen smoothly as you sail to better and better health with each passing day.*

Recovery is jumpy. It may feel haphazard and random. It is full of surprises. But, embracing the positive thought of recovery makes all the difference in the world.

Holding the Unconscious Belief You Do Not Deserve Good Health

A third reason why people don't really succeed in feeling better and getting relief from their symptoms is that part of them believes they do not deserve to feel better for whatever reason. This irrational thought form can be driven by deeply unconscious forces that are rooted in experiences that took place long ago and far away.

The Challenge of Sustaining Positive Thoughts about Recovery

Why is it so difficult to sustain positive thoughts about recovery? When you talk about your intent to start feeling better, chances are others may be telling you

- With their looks
- With their silence
- With their own thoughts (which we can also take in although they are not expressed)

> *"Oh ---you're being unrealistic. Do you really think you are going to recover? Get realistic my friend. Accept where you're at. That is what it is all about – acceptance of the fate you have been dealt.*
>
> *The doctor told you what was going to happen. The doctor sees people with your problem all the time. They know what will happen. They know the true story. Get with it."*

Remove the Label. I recommend that you no longer "label" yourself as a person who has Parkinson's Disease. Simply remove yourself from identifying with this thought form web.

The term Parkinsons Disease describes a particular set of symptoms. What happens when you label yourself as someone who has "Parkinsons Disease"? You become entangled in a negative thought form template that packs little power.

Direct your attention to the symptoms. That is signaling system the body uses to tell us what it needs in order to heal.

The diagnosis of Parkinsons is made by medical doctors who have extensive training in the diagnosis and treatment of disease. Once a person is diagnosed, a medical doctor can legally prescribe medicines and treatments.

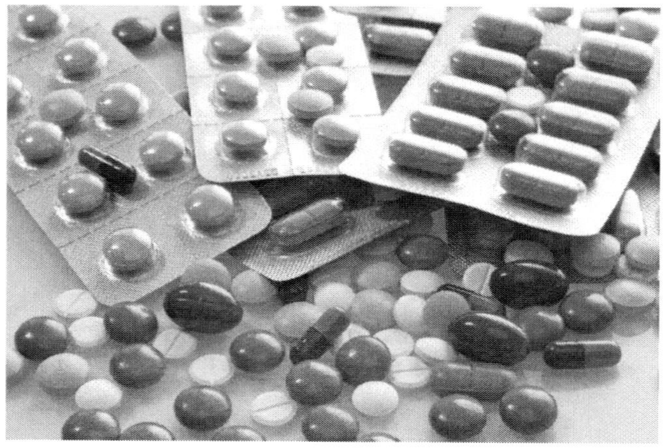

This approach obviously helps many people, but it is embroiled in a web of negativity. The focus is on what is wrong. The assumption is that the body is broken, but you do not have to adopt this negative thinking yourself.

The literature on Parkinsons presumes incorrectly that once the symptoms start, you are always going to get worse. Use your own common sense and intuition. Could it possibly be true that everyone is destined to get worse? Of course not!

Keep visualizations private. I also recommend that you keep your visualizations about recovery close to your chest. Keep them private until they have gathered sufficient energy to be materialized.

When we tell somebody else what we have been visualizing for ourselves – our dreams about the future, the desires of our heart – our visualizations lose their energy. Why?

We can see or sense judgment in the reaction of the other person. What are they really thinking as you are blabbing your innermost secrets?

> *"Oh, poor Robert. He really thinks that he is going to be able to ski again. He really thinks he is going to be able to work again. He really thinks he will be able to drive again. I love him, so I need to help him accept his fate."*

Your positive frame of thinking is instantly diluted. You shift into that negative energy of the wishy-washiness. Robert begins to think:

> *"Oh, maybe they are right. Maybe recovery really will not happen for me. Maybe I should resign myself to a sad fate of decline and deterioration. Maybe I cannot rejuvenate myself."*

The Key to Recovery

Be:

Clear

Direct

Focused

On track

Holding a positive intention will manifest recovery. You control what is going to happen. The image that you create for yourself in your mind's eye is the reality that you will manifest.

If you entertain wish-washy thinking, which is easy to do, I can guarantee you that you will manifest wishy-washy outcomes.

The chances are pretty good that your physical body will be flopping all over the place with one symptom after another. Symptoms shift swiftly with wishy-washy thinking. Symptoms this week will suddenly transform into another set of symptoms next week. That is simply the reality of holding onto wishy-washy thoughts about recovery.

It takes discipline to focus and stay on track with a positive intention to heal. It takes a strong will and decisiveness. Our thoughts always mold our future.

Reasons Why Recovery Waffles

Why do people with Parkinson's get worse? If you want to discredit your positive intention to get better you may be thinking,

> *"Ah, I just went to a support group. Talk about being depressing! Everybody was talking about ways to slow down the progression. They accept the fact they are going to get worse. They simply want to get worse a little bit slower. How much more negative than you get?*

The reason why so many people with Parkinson's do in fact get worse is because they believe and think that degeneration is inevitable. It is as simple as that.

How the Body Processes Negative Experiences

There is another reason it is challenging to maintain and sustain positive thoughts without entering into wishy-washy territory, the territory in your mind that repeats over and over,

> *"I am not going to be able to make recovery happen."*

Starting from the moment that you are conceived every time you hear anybody in your life

Criticize you

Attack you

Insult you

Judge you

Belittle you

The body soaks up the negativity. The negative energy is captured, stored and trapped in your body.

We all get a large dose of negativity in our lives, especially when we are children. It is chilling to observe how cruel children can be to each other. The cruelty of exclusion is hurtful.

We are all assaulted as adults by judgments from family members and friends who love us. We get cruel looks from strangers. All of that is soaked up by the electoral connections in our brain.

You may be thinking, you are getting a little on the fringe here. How in the world can this be the case?

There are 8 billion electrical magnetic contact points in our brain. Each contact point fires 3000 times every second. These tiny, microscopic events are barely detectable, but they can be measured with instruments.

The body encodes experiences in ways that are similar to how a computer stores information. Your brain stores everything - absolutely everything - that has ever happened in your life. We don't technically remember all those horrible events, but they are stored in our brains nonetheless.

There are billions and billions of negative thought forms that are literally stored in our brain. They are tiny. They are microscopic. They have a powerful influence on our lives.

When we begin to shift into that space of negative thought form recreations, we reactivate all of the contact points that we created long ago and far away. With eight billion contact points that are pulsing 3,000 times every second, the level of activity in the brain is unfathomable. The negative thought forms are churning all the time. It is easy for anybody to slip into a negative thought form template.

Having described and acknowledged the challenge posed by the billions of negative thought from electrical charges that are stored in our brains, it is time for the good news. Meditations can neutralize and remove the curse of the negative electrical charges in our brain which sustain us in a state of negativity.

Meditations

Faith Overview

Welcome to this meditation on the fourth step that is necessary for recovery from the symptoms of Parkinson's to unfold.

1. The first step - to focus.
2. The second step - to materialize.
3. The third step - to stay on track.

Now, the fourth step which in many ways bolsters and reinforces the importance of the first three.

4. The fourth step: Believe recovery will happen. Have faith.

At Parkinson's Recovery I often talk about "hope". The word "hope" is actually qualified. When I say the hope, there is both the possibility it may happen and there is a possibility it may not. Hope is the word I use that is the gateway in to recovering, the gateway to embracing the belief that yes,

- *I have the faith.*

- *I believe that I will recover from the symptoms that I am experiencing now.*

There has just recently in the last few days been announced a miraculous discovery of a microfocus spectroscopy. English doctors have done a marvelous piece of work which was announced in the United States in the grand city of Chicago.

This particular device uses tightly focused beams of x-rays to penetrate tissues throughout the body with great precision, but in particular the brain. The new device is able to identify metal eons, particularly iron metal eons. This is announced as a wonderful discovery for anybody with the symptoms of Parkinson's because they can get the test and then know whether or not they have pockets of iron in their brain. I laud this work in the work of all researchers and scientists. They are revolutionizing what we know about Parkinson's.

What I want to do in this meditation, however, is ask you to focus your attention on your reaction to such a discovery. Is your response to immediately say,

> *"Oh, I want to get that test I want to know what's wrong with me perhaps my symptoms are caused by iron in my brain. I am going to take some time and figure out how I can get this new and revolutionary test."*
>
> *"It will take me probably quite a bit of investigation and time to find out more about it. It's obviously new. I may have to travel long distance perhaps even to England. It will clearly*

> *cost quite a bit of money. I may have to sign off a number of waivers since this is clearly an experimental procedure but*
>
> - *I need to know*
> - *I want to know*
> - *I want to know the answer."*

Feel into the energy of that search for finding out what's wrong. It has a low-frequency.

> *What hope is embedded in all of that fever? The answer is none.*

So you get the test. You pay your $10,000. You sign the waivers and you get the result. You have iron in your brain.

Most people have iron in their bodies in their brain. So what? Now, what are you going to do about it?

Reflect back when you began to have symptoms. What was your response? Did you percolate into a fury of fear and anxiety investigating what in fact might be going on?

There is another response. That is the response of faith. Perhaps iron is an issue for you. What might be an alternative thought that you could hold? How about this one: The alternative thought is:

> *"Yea, it seems reasonable to me that iron might be the reason why I am having symptoms of Parkinson. Research*

certainly indicates there is a connection. I know what I'll do. I will give blood three times a year."

You see, giving blood purges out iron that is circulating throughout our entire system.

- *That's easy.*
- *That is hopeful.*
- *That is something positive to do.*

It doesn't take a lot of time or I might say investigation. It's easy simply to go and give blood. Make it so.

Notice how different the energy is behind the decision to take positive action and the decision to investigate a problem. The latter is negative energy and low-frequency. In contrast, the statement:

"I am going to give blood"

exudes positive energy to be sure.

- *You are doing something for someone else.*
- *You are giving back to the world.*
- *You are helping someone in a state of severe disease recover.*
- *You are helping yourself.*

That is easy, eh? Can you feel the difference between the two choices of how to think? Give blood. You know that you helped. That is all there is to it. It'll help others and by Jove it will help you. No way around it, there is positive energy here.

Believe it will happen is a key to staying on track and to focusing your intention on what you want to see happen.

> *If we are not clear about what we want ...*
> *If we are not clear about what we wish to see happen ...*

then we will manifest our wishy-washiness as you already know. Believe it will happen and it will happen. Do you then have faith or are you a doubter?

It is I assure you very easy to fall into the belief template of doubt and a nay saying.

I do admit and acknowledge I am one of the very few in the universe who says,

> **It can be done.**

What is your choice: Fall into the belief template of ***faith*** or fall in to the belief template of ***doubt***?

Even in saying this, notice the difference in the feeling that strikes your body.

Doubt is dark and dire.
Faith is positive and energizing.

The difference in the energetic charge of the words "doubt" and "faith" is huge indeed.

This meditation then is to help you investigate whether you really believe recovery will happen. Have you ever had faith in your life about something that needed to happen? Of course.

Everyone has faith. When you purchased a house - if you ever did purchase a house - and were at the table signing the papers, did you stop to read

- *Every word*
- *Every paragraph*
- *Every detail of the*
  - *Pages and*
  - *Pages of*
  - *Pages of papers*

you were committing yourself to honor legally? Perhaps so, but my guess is if you're like most people, you did not read all the papers. You trusted the lawyer who prepared the papers and the individuals responsible for setting up the possibility for you to purchase your home. They could be trusted. They were not trying to do you in. And you were right to trust them.

Where else have you had faith that something would happen in your life?

So, settle in.

Five Steps to Recovery

Get comfortable.

- Take a nice slow breath in and out.
- Place your focus on your breath, particularly as the air leaves your lungs.
- Fresh air in.
- Digested air out.
- Notice how each breath rejuvenates each and every cell of your body.

Focus your attention on what it feels like to let that precious air out of your body in to the universe. Feel your chest rise ever so gently and recede in. Listen to the sound of the air as it passes over your teeth, in and out.

With focused intention on just your breath, direct your attention to your body and relax whatever tension might be present in:

Muscles or
Tissues or
Tendons or
Organs

You simply say to the place in your body that is tight or anxious

"Not today honey. We are going to do it a different way today"

Focusing your attention on your breath and being aware of your surroundings with

- *the sounds and*
- *the smells and*
- *the feelings, the precious feelings of occupying the body that you live inside now.*

In this state of relaxation investigate for yourself now your hesitation and your own doubts about your future.

Do you carry with you in your tissues and in your thoughts doubt about whether recovery can happen or not, about whether you can in fact feel better from the symptoms of Parkinson's? Why might that be so?

- *Is there some draw or attraction to the fear that arises when you simply say the word Parkinson's?*
- *Is there some charge that you get by having these symptoms?*
- *Are you getting something out of it?*

  - Perhaps you seek the attention of others who you love?

  - Perhaps you are one who has always given to others through your work, through your volunteering, through your creativity.

  - Perhaps you needed an excuse, a reason why someone else had to at last give to you?

Think about it. There is no reason why you must have symptoms that are troubling in order for someone else to give to you now. You get that by asking for it. It's as simple as that.

You get that by

Being focused,
Materializing your wish,
Vocalizing it and having the faith that it will become true.

We all carry doubt. We all carry a "no" somewhere either consciously or unconsciously. Some of us have resistances to coming into our full power and life force. We push back our own inner creativity for difficult and complex and perplexing reasons that only the very sophisticated of psychologists and mental health counselors can figure out.

What does it matter? Simply acknowledge now to yourself and make conscious

Whatever hesitations
Whatever doubts
Whatever disbelieves you may be holding onto.

It does not matter what anyone else has ever said about the future for you. The only person who determines your future is you and your thoughts about the future.

Have you ever heard yourself say:

"Oh. I am just old."

There it is. If you are just old then,

"Yeah, I guess that's what happens to old people."

Holding onto that belief, carrying that thought, honoring it, giving it energy will ensure that your symptoms will be sustained and nurtured. You give them fertilizer. You nourish your symptoms by such thinking.

Have the faith that recovery will happen and it will happen for you in just the way that you materialize it to be so.

- *Perhaps not in this speedy time that you would like to see ...*
- *Perhaps not even in the way that you would like to see unfold ...*

Many people believe that if they have the faith

Every day will be better
Every day they will feel better

That's not true for anyone. I don't care what the condition of their health is. We all, each and every one of us, have bad days. We all on some days and even some weeks and even some months feel absolutely lousy.

If we attach a negative thought to that feeling, a thought like:

"Oh my god I'm going to die."

It will likely get worse. The secret is out. We are all going to die. What's the big deal?

Why attach a negative frame to that thought. It's the way life is supposed to be. Have the faith recovery will happen and it will unfold for you in its own time and in its own way.

- *Go back now to focusing.*
- *Go back now to vividly imagizing what you want to see happen for yourself.*
- *Hold now the faith that it will happen for you*

    - Set up your TV screen
    - Get out your controller
    - Turn the movie on
    - Watch your future on the channel that you have selected for yourself
    - Have the faith
    - It will happen for you in just the way you envision it on your television set now.
    - Stay on the same channel.
    - Colorize your picture
    - Detail it out.
    - Make it a movie picture of your life as you would like it to be tomorrow. Feel the joy of what it is like to be the new you.
    - Feel the surge of energy because there are so many things that are now possible for you.
    - Embrace the belief that anything is possible.
    - Have the faith.

No need to do any of this faith strengthening alone. Call in whoever needs to be present now with you in your imagination to support your commitment and determination to feel better. There is support for you each moment of your life.

- *Support from your parents whether they are living or not living.*
- *Support from all the persons in your life who you have loved whether they are living or not living.*

This is not about any particular religion, although your own religious beliefs may help and support your commitment to recover now.
Call on those beliefs now to strengthen and support your resolve that recovery for you will unfold.

- No doubts.
- No waffling
- No hesitation

Know in each and every cell of your body it will happen precisely the way you picture it to happen in its own time and in its own way. The universe returns to us precisely what it is that we set as our intention

- *Focus*
- *Materialize*
- *Discipline yourself and your thoughts to stay on track*
- *Be clear about what you want*
- *Have the faith that it will happen*
- *Trust*

Five Steps to Recovery

We are not talking miracles here. We are talking the reality of how thought forms work. Thought is the greatest power of all in the universe.

Your thoughts determine your future.

In your movie that you are watching now of the future, accept all of the pleasure that your future now offers to you. Embrace the joy of a new life, a new way of being.

Welcome the new creativity that will inevitably emerge if you allow it to be discovered. For you see, there are parts of you that you still need to become acquainted with as your life force becomes re-energized, not because of any particular therapy you elected to pursue - whether conventional or alternative -- but because your thoughts about all possibilities are

- *Clear*
- *Unwavering*
- *Dedicated*
- *Focused and*
- *Determined.*

Feel the lift of energy when you have faith. Feel the power behind knowing what will happen for you in your future.

See it.
Believe it
Have the faith baby.

It will happen for you.

Activate your faith. Sustain it now and for the rest of your life. Welcome the new you, the new life force.

Know that as you stay on track with each passing day, new revelations will unfold, new discoveries will be encountered. You will be giving a great gift to yourself and all of those in the world who you affect.

Know that you are on the road to recovery.
It will happen.
Have the faith.
Keep the faith
Always

Remove the Negativity

Settle in now. Bring the focus of your attention to your body. Relax whatever might be tense in your body right now as you move up from your body from your

- *Toes to your*
- *Thighs to your*
- *Belly to your*
- *Shoulders to your*
- *Arms to your head*

A quick scan will do. Something quick is all that you need. If you feel any particular muscle that's tighter than it needs to be, just loosen it up so that you can really become relaxed. The more relaxed you are, the more effective this particular meditation will be for you.

Take seven slow breaths.

- *Breathe in.*
- *Breathe out.*
- *Focus your attention as you breathe out*
- *Notice what it feels like as the air exits your lungs with each breath*
- *In and out*
- *Focus your attention on relaxing each and every cell and tissue and muscle in your body.*

I invite you now to think of a person who is living or who is not living to assist you with this activity. Think of someone who has helped you in the past, who was always been there for you in the past. Ask them to be physically present with you now. They need not be physically present. Just have them be present for you in your imagination.

For those of you who are deep into spiritual roots, I invite you to call in any spiritual beings that you would like to be present. For some of you this may consist of angels. When you ask them to help you out, they will come whether you believe in the existence of angels or not. So, why not ask?

It is the duty of angels to be present and to help when they are summoned. Whether you believe in angels or not, I invite you to call whatever angel you might want to call in. If you don't know the names of any angels, make them up or simply request the angel who needs to be present whatever the name or qualifications.

- *Call in people who are real for you and made a positive difference in your life*
- *Ask them to be present*
- *Don't do this by yourself*
- *Get help by simply asking for help.*

Imagine a world that contains imaginary magnets for the taking. Choose a magnet now. Your magnet can have whatever shape you would like for it to have, for the choices are infinite.

This magnet has a special capability. It can suck out negative thought form filament from your body.

See the magnet in your mind's eye

Sense it

Feel it

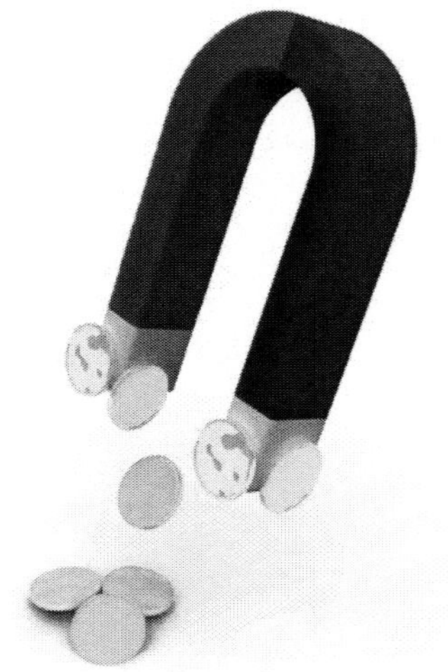

Place your magnet a couple of inches below your belly button. Stick it over what is called the second chakra.

Your magnet has and "on-off" switch. Turn the switch to "on" now.

Watch your magnet attract all of those negative thought form filaments that you have taken on over the course of your life. Allow the filaments to fix themselves to the magnet.

This is the place in your body where you store thoughts and feelings about yourself and others. Words spoken to you out of cruelty will have energetic chargers in this place on your body. There are billions of magnetic contact points here.

- *Invite the magnet to suck up all of the negative thought forms.*

- *Watch those filaments just as you might see lead filaments clinging to a magnet.*
- *Watch those filaments be attracted to the magnet that you have constructed,*
    - *All of the negative thought forms*
    - *All of the hurtful words*
    - *All of the harsh judgments*
    - *All of the cruelties that you have absorbed in your body*

Now, as this magnet becomes full

1. *Lift it out with your imagination from your belly.*
2. *Hold it up with your imaginary hand into the air.*
3. *Ask whoever you have requested to help to take all of those negative thought forms and dissolve them.*
4. *See them being released out into the ethers.*
5. *Invite them to be dissolved and released.*

With the release your magnet regains purity, a revitalized color, a renewed vibrancy and a reenergized life force that has a much, much higher frequency.

Take this very same magnet that is now been revitalized and re-energized and, with your imagination, insert the magnet where your stomach is located at the bottom of your rib cage.

This is the place in your physical body where you hold thoughts about your place in the universe, why you are here, what you are here to do and to accomplish in this lifetime. Allow your magnet again to attract all of those negative thoughts that have been absorbed through all of the billions of magnetic contact points in that part of your body.

Negative thought form filaments are being extracted from all parts of your body now, including your brain.

- *You can actually hear the filaments connecting to the magnet if you listen closely.*
- *You can actually smell a slight electrical smell as you might smell when there's an electrical short-circuit in an electrical wire.*
- *Smell the magnet soaking up all of these negative thought forms that you are carrying and have taken on throughout your entire lifetime.*

Is your magnet full now?

1. *With your imagination take your magnet out of your middle part of your body*
2. *Hold your magnet out into thin air.*
3. *Invite whoever you have invited to be present with you today to help you release all of these negative thoughts into the ethers of the universe*
4. *Transform those negative thought forms into neutral charges that can no longer affect you.*
5. *Watch now how the magnet revitalizes itself*
6. *Watch the colors brighten*
7. *Sense the power of the energetic charge in your magnet*

8. *Feel the support of those who are here present with you to take all of that out and release it*

Take the same magnet and wherever you're called to in your body, whether it is your brain or your feet, insert it in the place in your body that may still have some of those negative charges, those negative thought forms.

- *Comb your body with the magnet as you hold it.*
- *Allow the magnet to draw out all of the negative thought forms.*
- *Hear them clink to your magnet.*
- *Smell the electrical short-circuit charges*
- *Scan across all of your body where ever those negative thought forms may reside, sometimes in the throat for we don't always speak truths to others out of fear and hesitation*
- *Allow all the filaments to be absorbed into your imaginary magnet.*

For a third time, hold that magnet as those who were present with you now release all of those negative thoughts into the ethers of the universe. See those filaments release and fly away. The magnet recharges instantly, ready to take away at any request any negative thought form that you are holding in your body that may still reside there.

The magnet is always available to you. It is always there.

Beings living or not living that have been present with you in this meditation are always available to be summoned to help.

Take the magnet now and store it away in your body wherever you would like to store it. It will always be there for your use whenever you would like to activate it.

I invite you now to come out of your meditative state and become aware of your surroundings

- *Walls in the room*
- *Sounds*
- *Colors*
- *Smells*

Take a nice deep slow breath. Congratulate yourself on setting the intention to release the billions and billions and billions and billions of negative thought forms that you have absorbed throughout your entire lifetime.

Why Bother Neutralizing the Negative Thoughts?

Why is this meditation so important? There are so many negative thought forms that are stored magnetically and electrically in our body that the minute one negative thought pops into our head, it triggers a cascade of other negative thoughts. It is like a chain reaction. Once engaged it's hard to turn off and convert back over to the positive thought form mindset.

It is not that it is bad or irreversible when this happens. It happens to all of us, all the time. It is just important to consciously acknowledge when it happens.

When you float into that seductive space of negative thinking, take out your magnet (it is always handy) and begin to absorb all those negative thought forms that are spinning out of control in your mind. The magnet meditation can always help you get back on track no matter what you are doing.

The positive thought forms are what will make the difference between being able to recover or not recover. Wishy-washy thoughts coupled with negative thought forms will result in exactly the outcome you do not want to see happen: wishy-washy outcomes.

- *With courage*
- *With determination*
- *With focus*
- *With positive intention*

    - *Hold the positive thought and intention of what it is that you have set out to accomplish for yourself.*
    - *Ask is this what I really want?*
    - *Materialize it so that you can see exactly what it looks like.*
    - *Make it palatable.*
    - *See all the colors of it.*
    - *See all the people who are with you doing it.*
    - *Enjoy it.*
    - *Take immense pleasure in it.*

The more often you can:

Feel it

See it

Want it in your heart

The quicker it will happen

It is Not That Simple

There are many books and CDs that are proliferating the marketplace that talk about manifestation and being able to accomplish the goals that you set for yourself in your life.

My impression of much of that work is that it sounds all very simple. All you really need to do is a

1

2

3

4

Step process. The person who is either recording the audios or writing the book has obviously done it. What they say is

It is easy.

What I want to say at the outset of this week's series of meditations is that,

It is not easy to manifest recovery.

- It is not easy to stay on track.
- It is not easy for me.
- I don't think it's easy for anyone.
- It is hard work.
- Possible to be sure
- Doable to be sure

- With focus and intent you can be sure that it's possible, but it is not easy.

I want to say at the outset:

- Acknowledge the work that you have undertaken to pursue.
- Acknowledge to yourself and honor the importance of the work.
- Know that there will be challenges
- There will be difficulties
- There will be frustrations

It is inevitable, but the rewards will be worth it.

Four Memories from Childhood

Think back on your childhood. At the top of your head bring to mind four memories about being a kid and growing up in your family. Any four memories will do. The first four you think about will be the right memories to work with.

Are these memories good memories or bad memories? Are these memories of positive exhilarating experiences or are they memories of sad, unfortunate and unhappy experiences?

It is possible that among your list of four memories, the number of negative experiences and memories actually outweighs the number of positive. Not necessarily, but for many people they will often remember the unpleasant far more immediately and quicker than they will remember the pleasant experiences.

Why is this? The negative memories exist because when those events happened they tend to be on the resolved experiences that had no closure. These negative experiences get stuck as energetic magnetic charges in our bodies.

As it turns out many people had many pleasurable experiences growing up, but when they reflect on their childhood what they tend to think about are

- the sad times,
- the unpleasant memories,
- the scary events,
- the times when they were afraid.

On balance the truth may be that there were more pleasurable and happy experiences than there were negative and unpleasant experiences for the reason that I just described. The negative experiences tend to get stuck in our bodies. We have

billions and
billions and
billions and
billions

of them stuck as energetic, magnetic charges in our body. That is why when we have just one negative thought that happens to creep in during the day, what follows are a whole series of follow-up negative thoughts. One negative thought is an invitation to the billions of others to be activated.

Acknowledge that to be true. Simply acknowledge it today. As you think back on your childhood I invite you to remember four pleasant, happy, memorable experiences, times when you were truly and genuinely joyful and happy. Those events did occur they are just not remembered as readily as those awful horrendous events.

Pleasure Behind Pain and Suffering

Why is it so difficult to shed ourselves of having negative, unpleasant thoughts, negative thoughts that have low frequency and have no prayer of going anywhere other than to wishy-washy land? Why is it so tough to just declare:

> *"I'm not going to have negative thoughts anymore. I am going to be positive and have only positive thoughts."*

Here's the answer. When we are children our birthright is to have pleasure. Often times we are in circumstances as children when we are not in pleasurable circumstances. They may be:

- *Circumstances that are oppressive*
- *Circumstances that are hurtful*
- *Circumstances that constrain us in ways that are oppressive*

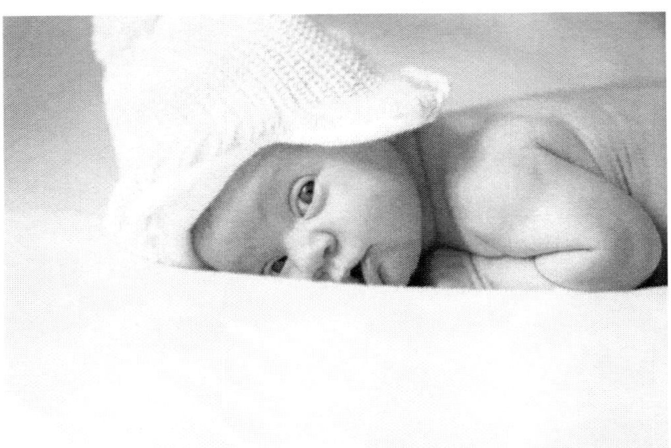

We cannot get out of the circumstances because we were children and we are dependent on those who care for us. Everything we are doing in this period of time may be a disaster. It may be hell on wheels. It may be a horror of all horrors in a child's mind's eye.

What does a child do in those circumstances where there is no positive pleasure to be experienced? The answer is:

The child attaches pleasure to the pain.

To a certain extent we all do this. Because pleasure is our birthright as a child, we will find pleasure in anything, even when it is horrible.

- *It is, you see, why some people become masochistic.*
- *It is why some people hurt themselves because actually that's pleasurable.*
- *It is why we see so many* aberrant behaviors in people.

The simple explanation is there is the attachment of pleasure to the pain. That is why we all sometimes unconsciously hold on to and treasure negative thoughts.

The meditation I have for you today is this. Think about and meditate over whether you are actually getting some pleasure - and it is negative pleasure and it is unconscious - but see if you can't feel a nudge of pleasure out of the agony and suffering and pain that you are suffering now. That attachment of pleasure is one of the reasons why it is so, so difficult to simply declare to yourself in your mind,

"I am now going to be positive"

That means you are giving up pleasure.

Energetic Charge of Positive Thoughts

Go back over the last few hours of your day-to-day. Remember a negative thought that you had whatever it might be. It does not necessarily have to be a huge negative judgment, just one recent negative thought.

- *Remember it now.*
- *Remember the substance.*
- *Remember the content however silly it might be.*
- *Remember it in your head*

Now, remember that negative thought in your body. What does your body feel like?

- *How heavy is your body as you rethink that negative thought.*
- *How sluggish do you feel?*
- *How slow and lethargic are you feeling as you think this negative thought?*

Shift now and think about a positive thought that you've had over the last several hours - whatever it might be - however inconsequential. Remember the content of that thought - something that happened recently - so it's very vivid in your memory.

Holding that positive thought I ask you now,

- *How does your body feel?*
- *How light is your body?*

- *What is the feeling?*

Compare the energy between the two thoughts (the positive and the negative). Contrast what it feels like holding the positive thought with what it feels like holding the negative thought.

The two have dramatically different frequencies and different energetic charges. The positive thought form can take you anywhere your heart desires. The negative thought form will take you to the cellar and to the sewer. The negative thought form has low energy and low frequency and little energetic charge. There's a huge difference between the two.

You'll get a whole lot more bang for your buck with the positive thought form than you will with the negative. Acknowledge the difference in your body. Learn to recognize the difference. The energetic difference between the two thought forms is huge.

Purge Negative Thought Forms from Your Body

Place in your hands now an imaginary magnet that is red in color. The magnet is the size of the palm of your hand and it is colored red. This magnet is a negative thought form collector which has two settings.

On one setting the magnet sucks up negative thought forms. With the other setting it repels negative thought form.

Take this imaginary magnet in your hand. Move it up to your head. Turn the switch to "suck-em up" and scan your entire head as this imaginary magnet sucks up the

- *Billions and*
- *Billions and*
- *Billions of*

negative thoughts that are stored in the magnetic electrical connectors in your brain. Physically move the magnet all around your head,

- *Top*
- *Sides*
- *Front*
- *Back*

Suck up all of those negative magnetic electrical charges. These are physical realities that are residing in your brain. You can exorcize them from your brain using your magnet.

Now, suck-em up. They look like little small iron filaments that link to clink to the magnet. If you listen carefully you can hear the clicking of these millions and millions of negative thought forms as they attach to your powerful "suck-em" thought form magnet collector.

- *Smell the electrical smell as if there is short-circuiting going on.*
- *Taste the bitterness.*

Move the magnet all around your head as you collect up all of those negative thought form charges. Now, take your magnet and hold it up high over your head to the sky. Switch the switch to repel so that all of those magnetic filaments are sent up to the universal ethers to be

- *Discharged and*
- *Dispelled and*
- *Dissolved and*
- *Forgotten forever.*

Free at last.
Free at last.
Thank God almighty, I am free at last.

Step Four: Have Faith

The fourth step has nothing to do with

- *Eating right or exercising correctly*
- *Going to this medical doctor or consulting with that naturopath*
- *Receiving treatments from this acupuncturist or that body worker*

The fourth Step to Recovery addresses the most fundamental piece of all, the piece that is absolutely essential for you to jump start your program of recovery so you will begin seeing relief from your symptoms. The coherent theme of the five steps is that "right thinking" leads to recovery

It is all about your thoughts. Thoughts are what will make a huge difference. Perhaps you see a good friend you haven't seen for five years. You were dear friends five years ago. You embrace. Your friend says to you

> *"How are you? How have you been?"*

With your head hanging low to the ground, you say:

> *"Oh, I have Parkinson's Disease."*

Feel in now to that energy. For me, this answer is depressing. It has a very low frequency. It is certainly not an uplifting statement. It may be factual from the perspective of a diagnosis from a medical doctor, but the fact remains that this greeting to a new friend is full of dark energy.

Now, imagine that you confront the very same circumstance. You meet a friend you haven't seen for five years. Your friend says:

> *"How are you? How have you been?"*

And, holding your head proudly and assertively, you say:

> *"Oh, I have some symptoms here and there, but I am well on the road to recovery."*

Check in with that energy. That energy is the energy that is full of hope. This thought form has the punch behind it which is necessary for recovery.

What then is the fourth step? The fourth step is conveyed by a very simple word. That word is faith. Those who have faith that recovery will happen for them will find that it will happen.

In much of my writing I talk about hope. I say to people:

> *Parkinsons Recovery is a place where you can find hope.*

And, this is certainly true. I document the stories of people who are feeling better and of people who are symptom-free today. I use the word hope frequently and liberally.

The truth is, however, that hope is a qualified term. When I use the word "hope" here is what we think to ourselves.

> *"Okay, there is hope. That is pretty great. It means that it might happen or it might not happen. I do not know for certain, but I'm hopeful there's a chance."*

What is the chance estimate in your thoughts? Is it 40%? What does it matter you say to yourself, at least I have hope.

As it turns out, hope is not the best word for me to use. If you are determined to see a relief from the symptoms of Parkinson's, the word to embrace is "faith." Faith is what will make a huge different and a compelling difference with regard to your ability to recover.

When Have You had Faith in Your Life?

Regardless of whether you hold deep religious convictions are not, you have had faith throughout your entire life. When you are driving and come to a red light do you not stop? Do you not count on the fact that when the light turns green, you can proceed safely with caution? You have faith that other people at that very same stop light will obey the rules of the road. You do. I do. Everybody has faith that the system of stop lights will work. It does work.

When you obtain a prescription from a pharmacist for medication that has been prescribed by your doctor, do you doubt that the medicine has integrity, that the medicine is in fact what you ordered and paid for? Are you a person who takes that medication and submits it to a chemical analysis to ensure that the medicine you received from the pharmacist is in fact what you bought?

Well, maybe so but I'm here to tell you I have certainly never done that and I do not know anybody who has. When we get a prescription from a pharmacist we have faith that pharmacist is well trained and qualified to fill the prescription. We have faith that they will provide us with the medication that we have ordered. We have faith that it will be just so.

Do you use an accountant to do your taxes? I've recently done that because the taxes got a bit complicated. Do you (if you use an accountant) have a second accountant that checks on the work of the first accountant?

Well, maybe some of you do, but I suspect most people don't. You have one accountant. The accountant does the work. You sign it off. You mail it off. The work is done. You have faith that your accountant knows what they are doing. Everybody does.

We have had faith in many, many decisions throughout our lives, day in and day out. What does faith mean to you?

- *Faith means that you believe it will happen. You believe that it is true.*
- *Faith produces confidence.*
- *Faith generates action. When we have faith we are willing to do something about it.*
- *Faith is a surge of power and energy that can heal every tissue and cell of our body.*
- *Faith encourages us to focus.*
- *Faith enhances our creative energies and juices so that we are willing to find other ways of making something happen for ourselves.*

- *Faith makes it possible for us to have the dedication to look for solutions in the oddest places.*

Think about it. When you have faith that something's going to happen don't you begin to attribute significance to even the most inconsequential events? For example, you make a trip on a plane and you sit next to somebody. That stranger begins to talk about a topic that is a turns out you're desperately interested in? If you have faith don't you think:

I wonder why that person is sitting next to me?

Okay, it might be entirely random, but the fact is that your interest is piqued in sitting next to that stranger. Things begin to happen for us in our life when we have faith. We are more willing to ask ourselves the curious question

- *I wonder why that just happened to me?*
- *Is that a signal?*
- *Is that a sign that I need to pay attention to?*

Faith means that you will find a way to make it happen. You will act without any hesitation whatsoever. So, many people wonder:

"Well, okay, how can I really have faith? After all there are so many naysayers. There are so many doubters. There is so much information that it is just not possible for me to begin to feel better from the symptoms of Parkinson's. How in the world am I supposed to have faith in something that even the most well trained and skilled people tell me is not possible?"

I believe that one of the ways you can have faith is simply to act. That is what you do. You say alright

"I am going to make it so."

Have faith that it is true and using the strength of that thought, take action.

What action do you take? I honestly think it actually doesn't matter what action you take. What matters is that you begin to take action. You begin to move. Maybe what you decide to do today is not working. Okay. You take more action. Action gives the energy needed to take another step. One step leads to two to three to four to five. Action is what can make a huge and profound difference.

Here is a question that I received that speaks to the topic of faith.

> *"Do you have any suggestions on how to get an 80-year-old man to recover from Parkinson's with extreme rigidity and dementia who is virtually at the non-responsive level? He can barely speak and has a very difficult time swallowing and is now developing aspiration pneumonia. Is there anything that could provide immediate relief?"*

Now the initial answer is:

> *"Yes, there is a lot that could be done that would provide immediate relief."*

The underlying issue with this particular individual however turns on a quite different question:

1. Do they really want to heal?
2. Do they really want to recover?
3. Do they have faith that it can happen?

What's really underneath this person's behavior? What I feel when I read this question is a severe state of depression. It feels to me like there is almost an underlying wish to check out. That place comes to all of us in our lives:

> *We live*
> *We do*
> *We love*
> *We have joy*
> *We have times that aren't so good*

Then, we get to a point where we say.

I'm done. I have done all I want to do in this life.

At this point, people check out. They stop taking care of themselves. Their symptoms become increasingly worse.

Health is an active process. It is what we all must do in order to maintain wellness. I am not referring to just people with the symptoms of Parkinson's. This is true of everybody.

It sounds like this particular individual is not taking any steps to take care of their own health. Has he given up all faith? Does this particular individual have no reason to stay here on the earth any longer? The question I have directly for this individual is:

> *"What do you have left to do here on this earth?"*

This 80 year old man is losing interest in eating and always wants to sleep. He can barely keep his eyes open. There it is. It may be that this person has come to the end of their ropes. They have no more faith that they will be able to feel better. Their thoughts are replete with negative energy. They have made their decision to check out.

This question wasn't written by this 80 year old man. It was written by a loved one, somebody who deeply cares about this person. The person who wrote the question is in a particularly challenging spot.

They want their loved one to recover and feel better. It is miserable to be a caretaker and watch the one that you love go downhill slowly. But, the most loving thing we can do for a person who's made a decision to basically check out of living in a physical body is to honor their choice. Who are we to make a judgment one way or another about how they decide to live their lives?

There is a second thing I want to recommend to the person who wrote this very profound statement. The one positive action you can take is to begin taking good care of yourself. In other words, do for yourself what you are hoping your loved one will do for themselves.

This means you must have faith that whatever symptoms you might have (though they are not Parkinson's) can be addressed through your own healing response. You take positive action. You have faith that yes whatever ailment you might have can be addressed one way or another. Honoring the thoughts of faith creates an incredibly positive effect on the person that you are so desperately concerned about.

The real issue is for you is: how can you accept this? How do you handle this? My suggestion would be to honor their decision, support their decision whatever it may be. Simply be present for them. Love them for all that they have been and for all that they will become. Let it unfold as it might.

I can assure you when you ask the question

> *Is there something that can be done?*

The answer unequivocally is

> *"Yes. There is a great deal that can be done to jumpstart a person who is in this particular place. The underlying issue is clearly the issue of depression, so the other question turns on what really lies under the depression for this person? What is causing it?"*

Through discipline, practice and meditations, we can transform our thoughts about recovery and strengthen our faith that it will happen. That is the purpose of the following meditation. May your faith and determination and focus strengthen day by day.

Meditations

Faith Overview

Welcome to this meditation on the fourth step that is necessary for recovery from the symptoms of Parkinson's to unfold.

1. The first step - to focus.
2. The second step - to materialize.
3. The third step - to stay on track.

Now, the fourth step which in many ways bolsters and reinforces the importance of the first three.

4. The fourth step: Believe recovery will happen. Have faith.

At Parkinson's Recovery I often talk about "hope". The word "hope" is actually qualified. When I say the hope, there is both the possibility it may happen and there is a possibility it may not. Hope is the word I use that is the gateway in to recovering, the gateway to embracing the belief that yes,

- *I have the faith.*

- *I believe that I will recover from the symptoms that I am experiencing now.*

There has just recently in the last few days been announced a miraculous discovery of a microfocus spectroscopy. English doctors have done a marvelous piece of work which was announced in the United States in the grand city of Chicago.

This particular device uses tightly focused beams of x-rays to penetrate tissues throughout the body with great precision, but in particular the brain. The new device is able to identify metal eons, particularly iron metal eons. This is announced as a wonderful discovery for anybody with the symptoms of Parkinson's because they can get the test and then know whether or not they have pockets of iron in their brain. I laud this work in the work of all researchers and scientists. They are revolutionizing what we know about Parkinson's.

What I want to do in this meditation, however, is ask you to focus your attention on your reaction to such a discovery. Is your response to immediately say,

> *"Oh, I want to get that test I want to know what's wrong with me perhaps my symptoms are caused by iron in my brain. I am going to take some time and figure out how I can get this new and revolutionary test."*
>
> *"It will take me probably quite a bit of investigation and time to find out more about it. It's obviously new. I may have to travel long distance perhaps even to England. It will clearly*

cost quite a bit of money. I may have to sign off a number of waivers since this is clearly an experimental procedure but

- *I need to know*
- *I want to know*
- *I want to know the answer."*

Feel into the energy of that search for finding out what's wrong. It has a low-frequency.

What hope is embedded in all of that fever? The answer is none.

So you get the test. You pay your $10,000. You sign the waivers and you get the result. You have iron in your brain.

Most people have iron in their bodies in their brain. So what? Now, what are you going to do about it?

Reflect back when you began to have symptoms. What was your response? Did you percolate into a fury of fear and anxiety investigating what in fact might be going on?

There is another response. That is the response of faith. Perhaps iron is an issue for you. What might be an alternative thought that you could hold? How about this one: The alternative thought is:

"Yea, it seems reasonable to me that iron might be the reason why I am having symptoms of Parkinson. Research

certainly indicates there is a connection. I know what I'll do. I will give blood three times a year."

You see, giving blood purges out iron that is circulating throughout our entire system.

- *That's easy.*
- *That is hopeful.*
- *That is something positive to do.*

It doesn't take a lot of time or I might say investigation. It's easy simply to go and give blood. Make it so.

Notice how different the energy is behind the decision to take positive action and the decision to investigate a problem. The latter is negative energy and low-frequency. In contrast, the statement:

"I am going to give blood"

exudes positive energy to be sure.

- *You are doing something for someone else.*
- *You are giving back to the world.*
- *You are helping someone in a state of severe disease recover.*
- *You are helping yourself.*

That is easy, eh? Can you feel the difference between the two choices of how to think? Give blood. You know that you helped. That is all there is to it. It'll help others and by Jove it will help you. No way around it, there is positive energy here.

Believe it will happen is a key to staying on track and to focusing your intention on what you want to see happen.

If we are not clear about what we want ...
If we are not clear about what we wish to see happen ...

then we will manifest our wishy-washiness as you already know. Believe it will happen and it will happen. Do you then have faith or are you a doubter?

It is I assure you very easy to fall into the belief template of doubt and a nay saying.

I do admit and acknowledge I am one of the very few in the universe who says,

It can be done.

What is your choice: Fall into the belief template of ***faith*** or fall in to the belief template of ***doubt***?

Even in saying this, notice the difference in the feeling that strikes your body.

Doubt is dark and dire.
Faith is positive and energizing.

The difference in the energetic charge of the words "doubt" and "faith" is huge indeed.

This meditation then is to help you investigate whether you really believe recovery will happen. Have you ever had faith in your life about something that needed to happen? Of course.

Everyone has faith. When you purchased a house - if you ever did purchase a house - and were at the table signing the papers, did you stop to read

- *Every word*
- *Every paragraph*
- *Every detail of the*
    - *Pages and*
    - *Pages of*
    - *Pages of papers*

you were committing yourself to honor legally? Perhaps so, but my guess is if you're like most people, you did not read all the papers. You trusted the lawyer who prepared the papers and the individuals responsible for setting up the possibility for you to purchase your home. They could be trusted. They were not trying to do you in. And you were right to trust them.

Where else have you had faith that something would happen in your life?

So, settle in.

Five Steps to Recovery

Get comfortable.

- Take a nice slow breath in and out.
- Place your focus on your breath, particularly as the air leaves your lungs.
- Fresh air in.
- Digested air out.
- Notice how each breath rejuvenates each and every cell of your body.

Focus your attention on what it feels like to let that precious air out of your body in to the universe. Feel your chest rise ever so gently and recede in. Listen to the sound of the air as it passes over your teeth, in and out.

With focused intention on just your breath, direct your attention to your body and relax whatever tension might be present in:

Muscles or
Tissues or
Tendons or
Organs

You simply say to the place in your body that is tight or anxious

"Not today honey. We are going to do it a different way today"

Focusing your attention on your breath and being aware of your surroundings with

- *the sounds and*
- *the smells and*
- *the feelings, the precious feelings of occupying the body that you live inside now.*

In this state of relaxation investigate for yourself now your hesitation and your own doubts about your future.

Do you carry with you in your tissues and in your thoughts doubt about whether recovery can happen or not, about whether you can in fact feel better from the symptoms of Parkinson's? Why might that be so?

- *Is there some draw or attraction to the fear that arises when you simply say the word Parkinson's?*
- *Is there some charge that you get by having these symptoms?*
- *Are you getting something out of it?*

  Perhaps you seek the attention of others who you love?

  Perhaps you are one who has always given to others through your work, through your volunteering, through your creativity.

  Perhaps you needed an excuse, a reason why someone else had to at last give to you?

Think about it. There is no reason why you must have symptoms that are troubling in order for someone else to give to you now. You get that by asking for it. It's as simple as that.

You get that by

Being focused,
Materializing your wish,
Vocalizing it and having the faith that it will become true.

We all carry doubt. We all carry a "no" somewhere either consciously or unconsciously. Some of us have resistances to coming into our full power and life force. We push back our own inner creativity for difficult and complex and perplexing reasons that only the very sophisticated of psychologists and mental counselors can figure out.

What does it matter? Simply acknowledge now to yourself and make conscious

Whatever hesitations
Whatever doubts
Whatever disbelieves you may be holding onto.

It does not matter what anyone else has ever said about the future for you. The only person who determines your future is you and your thoughts about the future.

Have you ever heard yourself say:

"Oh. I am just old."

There it is. If you are just old then,

"Yeah, I guess that's what happens to old people."

Holding onto that belief, carrying that thought, honoring it, giving it energy will ensure that your symptoms will be sustained and nurtured. You give them fertilizer. You nourish your symptoms by such thinking.

Have the faith that recovery will happen and it will happen for you in just the way that you materialize it to be so.

- *Perhaps not in this speedy time that you would like to see ...*
- *Perhaps not even in the way that you would like to see unfold ...*

Many people believe that if they have the faith

Every day will be better
Every day they will feel better

That's not true for anyone. I don't care what the condition of their health is. We all, each and every one of us, have bad days. We all on some days and even some weeks and even some months feel absolutely lousy.

If we attach a negative thought to that feeling, a thought like:

"Oh my god I'm going to die."

It will likely get worse. The secret is out. We are all going to die. What's the big deal?

Why attach a negative frame to that thought. It's the way life is supposed to be. Have the faith recovery will happen and it will unfold for you in its own time and in its own way.

- *Go back now to focusing.*
- *Go back now to vividly imagizing what you want to see happen for yourself.*
- *Hold now the faith that it will happen for you*

    - Set up your TV screen
    - Get out your controller
    - Turn the movie on
    - Watch your future on the channel that you have selected for yourself
    - Have the faith
    - It will happen for you in just the way you envision it on your television set now.
    - Stay on the same channel.
    - Colorize your picture
    - Detail it out.
    - Make it a movie picture of your life as you would like it to be tomorrow. Feel the joy of what it is like to be the new you.
    - Feel the surge of energy because there are so many things that are now possible for you.
    - Embrace the belief that anything is possible.
    - Have the faith.

No need to do any of this faith strengthening alone. Call in whoever needs to be present now with you in your imagination to support your commitment and determination to feel better. There is support for you each moment of your life.

- *Support from your parents whether they are living or not living.*
- *Support from all the persons in your life who you have loved whether they are living or not living.*

This is not about any particular religion, although your own religious beliefs may help and support your commitment to recover now.
Call on those beliefs now to strengthen and support your resolve that recovery for you will unfold.

- No doubts.
- No waffling
- No hesitation

Know in each and every cell of your body it will happen precisely the way you picture it to happen in its own time and in its own way. The universe returns to us precisely what it is that we set as our intention

- *Focus*
- *Materialize*
- *Discipline yourself and your thoughts to stay on track*
- *Be clear about what you want*
- *Have the faith that it will happen*
- *Trust*

Five Steps to Recovery

We are not talking miracles here. We are talking the reality of how thought forms work. Thought is the greatest power of all in the universe.

Your thoughts determine your future.

In your movie that you are watching now of the future, accept all of the pleasure that your future now offers to you. Embrace the joy of a new life, a new way of being.

Welcome the new creativity that will inevitably emerge if you allow it to be discovered. For you see, there are parts of you that you still need to become acquainted with as your life force becomes re-energized, not because of any particular therapy you elected to pursue - whether conventional or alternative -- but because your thoughts about all possibilities are

- *Clear*
- *Unwavering*
- *Dedicated*
- *Focused and*
- *Determined.*

Feel the lift of energy when you have faith. Feel the power behind knowing what will happen for you in your future.

See it.
Believe it
Have the faith baby.

It will happen for you.

Activate your faith. Sustain it now and for the rest of your life. Welcome the new you, the new life force.

Know that as you stay on track with each passing day, new revelations will unfold, new discoveries will be encountered. You will be giving a great gift to yourself and all of those in the world who you affect.

Know that you are on the road to recovery.
It will happen.
Have the faith.
Keep the faith
Always

The Auction

I invite you over the next 15 minutes is to go into a meditative state. Go into a space of your own. Get comfortable wherever you might be. Make sure that your feet are on the floor. Take a breath in and out

- Feel all of the tension that might be residing inside the trunk of your body to dissipate
- Notice when the breath leaves your body
- Pay attention to what it feels like when the breath leaves your body
- Focus your thoughts now on your body and on your breath.
- Take another breath in
- Let the air out
- Relax each and every tense muscle and tissue in your body
- Make a quick scan to check
    - *your feet*
    - *your thighs*
    - *your trunk*
    - *your buttocks*
    - *your back*

Notice any tension there. Ask your body to relax now. There is no reason to be uptight. You are in a safe place. You are doing something useful. Let it go for now.

Allow all the tension throughout your entire body

In your shoulders
In your arms
In your neck
Around your eyes
Behind your eyes
In your ears
All the way up to the tippy top of your head

Do a quick scan. If you notice a little tension, just let it go. Take another breath in and another breath out. Treasure what it means to be alive. Treasure the joy of being able to breathe.

Today there is a very special event that you are invited to. This event is a thought auction. You may have never attended a thought auction before, so this promises to be a very special event indeed.

To prepare for this event, I invite you to bring whatever money you would like to bring because the thoughts will cost something. Retrieve some money, wherever you would like to take it from,

Whether it is on credit cards

Whether it is in cash

Whether it is stuffed underneath the mattress

Where ever the money is, stuff it into your pockets however much you want to bring with you to purchase these thoughts. Take whatever amount you think it might be worth to you today. Stuff the money into your pockets (obviously in your imagination).

How much money you really want to bring to the auction? How much is it worth to you?

Now, in your imagination, open up the door and enter into either your car or the trolley or the subway or however you get to places. Get into that mode of transportation.

Travel now to the site of the thought auction. It only takes you a few seconds to get there because we are in magic time now. You arrive at a huge, gorgeous auction hall. Enter through the front door. There are hundreds and hundreds of people of all shapes and colors and nationalities that are furiously anxious about being able to purchase the thoughts that they have come to this auction to purchase.

On the walls are gold vases. There are chandeliers made of such gorgeous glass. You have never seen the beauty of what it looks like when the sun shines through those crystals.

The chairs in this thought auction are incredibly comfortable. Go and sit in a chair. There are still a few that are available for you. Find a good seat so you can hear and see the auctioneer.

The chair seat gives you a massage. It vibrates. That is sweet. Talk about being relaxed, this is the coolest place to be of all today.

The sound of the auditorium begins to settle a bit now, a little bit quieter, as the auctioneer pounces to the very front of the room to stand behind an awesome golden podium. What an impressive hammer this particular auctioneer has for ending the bidding on each thought.

> *"Here. Here. "*

the auctioneer says.

It is time to begin the bidding. All persons of all colors and nationalities immediately sit in their seats. Each and every seat is now filled with the people who are here to purchase the thoughts that are so dear to them.

The auctioneer now starts

> *"I am now ready."*

Says the auctioneer

> *"to offer for purchase the first thought. The first thought up for purchase is*

"I can't do it by myself."

Low and behold, the person next to you says $1000 and the person to your right says $5000. People in the back bid $10, $20, $25, $30, $35, $40, $45. Finally, the auctioneer says

> *"Do I hear more?"*

no more.

"Sold for $45,000!"

You might have wanted that thought but I am sorry. That thought is already taken. Next thought for auction

"I'll never get better."

Curiously somebody from the very rear of the auditorium shouts $50,000! The person in front of them shouts $60,000, then

Seventy thousand
One hundred thousand
One hundred and fifty
Two hundred thousand
Two hundred and fifty thousand

Do I hear more than $250,000?

The auctioneer asks.

Sold!

Sorry. I do not know how much money you brought to this auction but that's not for sale. The third thought up for auction is

"I don't deserve it."

Somebody starts with $500,000. There are no other bidders. Do you want to buy that one?

"I don't deserve it."

It's up for sale. $501,000. You can have it.
The auctioneer says no, that is it. It is sold. I'm sorry. That is not one you can buy.

Next:

"I deserve to be sick."

Somebody - they're sitting just to the diagonal of you in the row in front – raises their hand and says 1 million. Everybody hushes up. The auctioneer says

"Sold. You can have that"

One more thought is going to be auctioned off and you haven't bought any yet. Maybe this is the one that you would like to purchase.

"Others are better off because I am sick."

Who wants to buy that? By Jove, there is somebody.

- *She is dressed in a golden fleece.*
- *She sits on the front row*
- *She stands proudly up*
- *she says* 100 million dollars

Without further bidding the auctioneer says with great pleasure and glee

"Sold"

The lights in the auditorium suddenly shut off. It's dark. It is a little difficult to find your way to the door. There is a light shining through the door at the entrance of the auction hall. You can almost see where you need to go amidst the crowd of people that are scurrying to leave.

Your attention is drawn to a sparkling clip chart at the door. Standing next to this clip chart is this oh-so-gorgeous woman. She is dressed in her natural clothes and seems to be lit up by the sunlight. There is a magical glow to this woman who stands next to this clip chart which has the word Free written at the top.

There are four statements on this flip chart

1. I can focus on what I want
2. I can envision it.
3. I can stay on track
4. I have a faith.

Four thoughts. Four positive thoughts that are packed with all the energy that is needed to recover from anything and surely recover from the symptoms of Parkinson's. She looks at you. She says,

> *"Here they are. They are free. You can take them if you want."*

I invite you to take those four thoughts for your own. As you take those thoughts in - those four positive thoughts which of course are the four steps to recovery - this most gorgeous woman who is the mother of the divine earth says to you:

By order of the divine universe you now own these four thoughts forever.

Know that they did not cost anything, Make them yours. Have faith that it is possible to recover. People are feeling better. People are recovering.

If you try to model exactly what somebody else does, the chances are it's not going to be exactly what you need. You have to find your own path. You have to find your own way.

At Parkinson's Recovery we know that there is no panacea. There is no one therapy or treatment or drug or approach that is going to work for everybody. It doesn't exist and it never will.

There are sets of approaches and therapies that are perfect for each and every person. How do you find what those are?

- *You find them when you have faith*
- *You find them when you have focus*
- *You find them when you're able to envision what it is that you truly and genuinely want in your life*
- *You find them when you stay on track*

If you see a guaranteed formula of success, there is no way you can fail.

1. Focus
2. Materialize what it is that you want
3. Stay on track
4. Keep the faith.

Five Steps to Recovery

Perhaps most importantly have faith. You have always had faith in the past. Every day you exhibit signs of faith. Use that now to your benefit.

- Embrace the belief that it is possible.
- Inculcate that into every tissue and muscle and cell of your body.
- When you know it is true in your heart it becomes true.

Our thoughts are the most powerful entity in the universe. When we think positive thoughts, positive things happen in our lives. When we take on negative thoughts that have very low frequencies, very negative things happen in our lives.

What is the choice for you?

- *Are you for some reason getting off on fear?*
- *Is it trippy to be ill and in some sense exciting?*
- *Are you getting a charge from being afraid?*
- *Are you addicted to anxiety?*

These are issues that have deep roots in your unconscious. If you set your intention they can be resolved. It is really then a choice, a choice to be made about how you want to live your life.

Everyone has issues. Everyone has symptoms of one character or another. No one is exempt. We all have

Issues that are emotional
Issues that are physical
Issues that are spiritual

It's just a reality of living in a body. So, what's the choice? Live in fear or live in love?

I invite you now to come up to greater consciousness. Emphasize and honor and enforce the power of positive thoughts. Our thoughts always affect our outcomes.

Summary

Listen to your thinking. If you discover you are wishy-washy and full of negative nay saying, it's a surefire prediction that you're not to be feeling well over the next several months.

If on the other hand you

Carry the faith of recovery
Know in your heart that it is possible to feel better
Hold a passion for doing what it is that you know you are here to do

Then, your life will be full of:

Joy and
Exhilaration and
Excitement and
Adventure and

Yes. You will be able to work through whatever symptoms you might have.

Power of Words

Language does no justice to being able to describe the power of thought. How many times have you told somebody else

"I have Parkinson's disease."

How many times? None? Perhaps you did not want anyone to know?

- *Five?*
- *Ten?*
- *Fifty?*
- *One hundred?*
- *One thousand?*
- *Ten thousand times?*

How many times? What did it feel like each time you told someone

"I have Parkinson's disease."

That has low energy. It is not going to get you to the place you want to be.

I invite you to use different words that are coupled with a different thought. When you talk with your

- *family and*
- *friends and*
- *coworkers*

If you wish other people to know you have the symptoms of Parkinson's disease then say instead

"I am recovering."

Feel the difference between the two.

I am recovering.
I am recovering.
I am recovering.
I am recovering.

- *Feel the difference.*
- *That has energy.*
- *That has power.*
- *That has promise.*

That will unfold in the way you want it to unfold. Try a different way. Try saying

I am recovering.

Pray Today

Whether you are religious or not, whether you are spiritual or not, I invite you to use a prayer for yourself today. A prayer, after all, is simply another way of having a positive thought form. The most powerful force in the universe is thoughts.

Instead of praying for others - your family, your friends, people who you don't know in the world who are suffering - use your prayers today for yourself.

What do you pray you wish to see happen for yourself? We are not talking about being selfish here. We're talking about acknowledging what you want to manifest in your life.

Pray for what you want.
Pray for what you wish.

You may have your eyes open or closed. You may do it alone or in a crowded store. You may pray anywhere throughout the day. Use positive thinking to pray for what you want to see happen for yourself.

- *No limitations*

- *No hesitation*
- *No waffling*
- *Take action*
- *Do it.*

It is the doing of it that makes it happen.

Pray.
Pray for yourself.

All of those in your life who love you will thank you for your prayers.

It is time today to attend to your own needs.

A Time When You had Faith

Bring to consciousness today ...

- A time in your life when you believed - when you had had faith that something would happen.

- A time when you believed that someone you loved would get well and they did get well.

- A time when you believed and had faith that something that needed to happen for you would happen and did happen.

Think back and allow that memory and that experience to float to the top of your consciousness,

A time when you had

Intense
Unequivocal
Unwavering
Focused
Concentrated belief that something would happen.

- A time when you had faith.

- A time when you held that in your body

- A time when there was

  no doubt
  no waffling
  no wishy-washy going back and forth

- A time when you knew in your soul and your body and in your heart that yes this was going to happen.

Invite that feeling to permeate each and every cell of your body.

- ***Receive it.***
- ***Accept it.***
- ***Make it a part of you.***

This is your life force.

It is there for the taking.

Take it for yourself now.

Muster Up the Times

It is time to muster up and acknowledge all of the times in your life from the moment you were conceived until this moment of your life when you had faith.

- *All of the moments*
- *All of the times*
- *All of the times*

when you trusted that a friend would do what he said he would do.

- The times when you went to a store to purchase a product and you trusted that product would work as it was advertised

- The times in your life when you trusted that a sacred partner would not betray you

- The time in your life when you trusted that you would get the job that you always wanted to get

Scan your life. Go from the beginning through the end and pick out all of those moments when you had faith. There are many. Pick them. Place them into a map that you can see in your mind's eye.

Notice how long this list is:

- *Reams and*
- *Reams and*

- *Reams of paper*
-

Filled with entries, entries in your life when you did have faith.

- *Pages and*
- *Pages and*
- *Pages and*
- *Pages*

of moments when you did have faith.

Cumulate now each and every one of those

- *thousands and*
- *thousands and*
- *thousands*

of moments when you had faith. Gather them together into one huge, magnificent mountain of faith. Climb that mountain step by step.

 Robert Rodgers, Ph.D.

It will take you

- *Many*
- *Many*
- *Many hours.*

Climb the mountain until you get to the top and raise your arms in victory. This you see is the power of faith that you carry now into your life.

Treasure it.
It is a gift you give to yourself.
It is free for the taking.

Five Steps to Recovery

Keep the Faith. Listen to Your Body.

I beg you today to have faith in one truth. The answers for you are within you. Your body knows what it needs. If you asked your body you will:

> *Hear or*
> *See or*
> *Feel or*
> *Sense the answer.*

The answers are not outside you. The answers are not from research universities or Parkinson's Recovery or any one sage or master or shaman or professor or wise guy. No. No. Have faith that your best decision can be to look within.

The wisdom of the universe and all that there is to be known is contained within each individual cell that resides in our body. Acknowledge the sanctity and sacredness of your body.
Your body is giving you information, precious information about what is up.

Listen
Ask

What is this all about?

Have faith that you can always trust your intuition. Trust the advice that you hear from your body.

Five Steps to Recovery

- *Your body knows the answers.*
- *Have faith.*
- *Know it to be true.*
- *You are all on the road to recovery.*
- *It will and is happening.*

Know it to be true as you listen to the messages your body gives to you today, moment by moment

- *Each twist*
- *Every pain*
- *Every nuance*
- *Every agony*
- *Every joy*
- *Every pleasure*
- *Every happiness*
- *Listen*

Your body is talking to you every moment. Do you feel abandoned by others? The best thing you can do is not to abandon yourself. Be there for you today.

Step Five: Persist

1. The first step of recovery is to focus
2. The second step of recovery is to detail out and materialize what you have set your heart's intention to accomplish
3. The third step of recovery process is to stay on track
4. The fourth step of recovery is to keep the faith, to know that in fact recovery will happen
5. The fifth step of recovery is the most important step of persistence.

We all have challenges with one or more of these five steps. They really are the key to manifesting anything in life. All five steps have to do with the right thinking. When our thoughts are in order, when our thoughts are clear, positive and focused we can manifest what we send our intention to manifest.

Everyone has difficulty with one of more of these steps. My personal challenge is the third, to stay on track. I suspect if you did some soul-searching, you will recognize one or more of these steps is a challenge for you.

What are my issues with staying on track? Whenever I initiate a project I am easily diverted into doing something else. Maintaining the discipline to stay on track is a true challenge indeed. I juggle five projects at the same time, diverting my attention from one to the other. It takes time for me to finish anything.

However, the fifth step of persistence is a step that I do up right. If you look at my own personal history or if you read a short biography that I might write about myself, it looks on the surface like everything has been easy in my life.

I've graduated from outstanding universities: an undergraduate degree from Vanderbilt University, a master's degree from Cornell University which is an Ivy League school and a PhD from a Big Ten university, Michigan State University. On its face it looks like my life has been

Snap

Snap

Snap

This guy had no problems getting through all of his formal educational programs. Anything but that is the truth. I could give you story after story of challenges I have confronted in just being able to complete my academic degrees.

I offer one example to give you a flavor of my life. One of the challenges of earning a PhD degree is passing a qualifying exam to write the dissertation. I finished all that was required of me in terms of course work, earning top grades in everything. I prepared for my qualifying exams to be advanced to candidacy so I could write my dissertation under the guidance of my committee.

I went before the examining committee and after a couple of hours of grueling questions they decided that I needed to have some more coursework and more formal background in one of the subject areas.

I said,

> *"Okay. That would be just fine. I am determined to be able to be awarded a PhD so that I can do research, so I will do whatever you say is required."*

I completed more coursework. I returned for my second examination. I appeared before the very same examining committee of professors. After a couple hours of more grueling questions, the committee decided that they remained unclear about whether or not they wanted to advance me to candidacy. It appeared as though I was still deficient in one or two of the subject areas they thought I needed to know more about.

I was determined. I didn't care who said what or who thought what. I was determined to earn my PhD from Michigan State University. I was determined to become a professor.

I went back to the committee several weeks later and said

> *"I'm ready to do this. I am qualified to be advanced. I insist on having a third examination now."*

They decided fine, we will go ahead and do a third examination. And after the third examination, the committee decided that yes, I was ready to be advanced to candidacy.

I've had many, many situations in my life that constituted huge disappointments and failures. When I actually became a professor at the University of Texas and then the University of Kentucky, time after time I would submit a journal article to what we would call perhaps a low prestige journal only to be turned down, to be told my article was not acceptable.

Using my knack of being able to persist, I would decide

> *"I'm not going to go a notch lower. I am going several notches higher."*

I would then submit the article to an even better journal, a class A journal. I would keep reworking the research article until I succeed in getting it accepted at the best journal. It would take years, but I pulled it off.

I was able to get my articles published not just in B journals but the best journals. I was awarded my PhD degree from Michigan State University. I was promoted to full professor with tenure.

I have failed time and time and time again in my life. There really is one factor that made the difference. That factor turns out to be persistence.

My advice to my own PhD students was simple. The one thing they needed to be able to get through their programs was to persist, to never quit. The issue was never ability. The issue was never creativity. The issue was never how much a person remembered.

The issue of completion always turned on whether they came to a point when they just gave up. The PhD students who became discouraged early on because of being asked to do something again obviously never finished. They quit. Those that were determined to finish their programs always did. It was always the case in my experience that any student who persisted was able to prevail in the end and be awarded a PhD degree.

What happens when you confront roadblocks and setbacks? Are you able to stick it through? The frustrations, disappointments and obstacles can be formidable. What happens to you when you are challenged with a setback? Are the medicines or therapies you are using today not working? Are you experimenting with others? Are you persisting?

Persistence is critical to recovery because setbacks are so common. Consider two examples of therapies where it is terribly easy to become discouraged and give up.

Detox Programs

The first general set of therapies where it is easy to become discouraged is with detoxing your body from toxins, heavy metals, pesticides and who knows what other noxious substances might be lurking inside.

What is the issue with anybody who pursues a detox problem? When you begin a detox program it is very likely that no matter what avenue you pursue, no matter how you're going about detoxing, you probably are going to begin to feel worse before you feel better. This is not always the case but when we encourage the body to remove the substances that are not supposed to be there, the body usually sends signals that it is struggling. Symptoms of detoxing can be

Nausea

Headache

Fatigue

When you start a detox program it may feel like you are actually getting worse. If you have decided that if you are going to start any therapy you have to feel better immediately, recovery is going to be a true challenge.

If you really want to never feel worse no matter what you try, the better alternative is to take medications. There are many great meditations on the market that numb you and that are specifically designed to mask symptoms. Medications can also have harmful side effects, but it is also possible that when you take them you will not feel worse. It is also possible that you will actually feel a little bit better, if not a lot better, in the short term. The long term implications of using some medications of course can make you feel worse. It depends.

To summarize, setbacks are common with any detox program. The tendency is to feel a little bit worse before you feel better. We always encourage people to proceed as tenderly and as gently as possible with any detox program so that uncomfortable side effects do not creep in and make life truly miserable for you.

Body Work

The second general area that can be a true challenge is to find ways to get relief from tension and stress and trauma from your physical body. Trauma and stress and tension are housed inside the physical cells. It can be a genuine challenge when we decide to confront the traumas and stress that we have experienced in our lives.

Five Steps to Recovery

When a person with Parkinsons decides to receive any of the body work therapies that can help release trauma, they may feel a great sense of relief afterword.

- *They may feel wonderful.*
- *They usually get a burst of energy.*
- *They often are able to do things that they have not been able to do recently.*
- *Their movements are more fluid.*
- *Their speech is louder and more assertive.*
- *Their attitudes become upbeat and positive.*
- *They are joyful and happy.*

These are all immediate outcomes that are possible from relieving stress and trauma.

A wide variety of therapies can be used to release stress and trauma. John Coleman, ND, recommends Bowen Therapy which is a gentle body therapy. Through Parkinson's Recovery, we offer clients craniosacral therapy which is a very light touch energetic approach which works with the body to unwind and release stress and tension. Craniosacral work follows the body to release the trauma that is trapped in the physical body.

The evidence of trauma is evident in the distortions of our physical bodies and in how we carry ourselves when we move. When you look closely at a person when they're standing, you often times will notice they are not standing up straight.

There can be twists in the trunk of their body.
There may be a side of the body that is reaching for the sky.
There may be a side of the body that is sinking down to the ground
They may be permanently slanted either to the left or to the right.
They may have one shoulder that is higher than the other.
Their face may have a lack of symmetry.
One side of their face may be more muscular than the other. Eyes may not be level to the ground.
A nose may be crunched to the left or to the right side.
One Ear may be higher than the other.

Physical abnormalities are a function of traumas that we have experienced in our lives. They may be created by accidents that created physical injuries. They can also be created by emotional traumas which can be the origin of significant trauma that is carried by the physical body.

With trauma, bodies get crunched up. They get wound up. That's one of the reasons why the neurological system gets so contaminated with toxins. Neural networks can't fire efficiently because there is a structural twisting of the body that short circuits the physical activation of neural networks which are very long and very sensitive.

This is why body work - regardless of which type of body work you might choose to pursue - can be so helpful and useful.

Bodywork is an invitation to release what tension and stress has been wrapped up physically in the tissues of the human body. The body literally begins to unwind.

A good way of illustrating what happens in the body is to think about what happens with a wash cloth when it is wet, twisted and then set aside to dry. Once dry, the wash rag transmutes into a solid piece of material.

That's precisely what happens in the body. Tissues that have been traumatized literally wind up just like a wet wash cloth. When dried over time they become set physically just as wet cement is set when it dries.

This is a significant obstruction to the efficient functioning of our intricate neural network system. When trauma is released from the body using any of the body work therapies, it can create a sense of discomfort emotionally. Emotions may come up that had not been felt before. I cannot report that the aftermath experience will necessarily be one of

Joy and
Bliss and
Happiness

When stress and trauma are trapped in our physical bodies it's important to release it. In so doing, the tissues in the body also relax. People can have memories that they haven't had for many years of events that occurred years ago. All of those experiences are brought up to be released. That is why the memories come up – to be released. The feelings are felt so that they will no longer be trapped in your tissues. The memories are present for you to acknowledge them so they can be released forever.

Memories that are rooted in incomplete, unresolved past experiences are connected to the traumas of our life. By uploading them from our unconscious and bringing them to the surface, the interrupted experiences can come to a full completion. Emotions wash through the tissues of our body. We release the trauma. We move on with our life.

When a person receives bodywork or craniosacral work over a several month period of time, their physical appearance also changes. A motivation I've always had for getting body work is that I look and feel younger. The more body work you get the younger you actually look.

I personally do not understand why people choose to get face lifts these days. Body work is completely safe, nonintrusive, far less expensive and much more effective in helping us look younger.

- If you choose to pursue body work ...

- If you decide your symptoms are being aggravated by a lot of stress and trauma that you have experienced,
- If you decide you would like to be able to release the traumas of your life so that you are no longer holding it in the tissues of your body
- If you decide to choose a path of healing where you're going to feel emotions – some of them unpleasant - that you haven't felt recently

Persistence will be required. The journey at the outset will not necessarily be a joy ride or pleasant.

The good news is that the unresolved traumas can clear through the tissues of your body quickly. Once the trauma resolves itself it is no longer a lingering curse that haunts you.

Releases will move you into a profound sense of elation, contentment and happiness. The body will begin to move much more fluidly. Pain will usually subside and become more tolerable, if not absent. There are results of bodywork that are quite marvelous to report, But they are not necessarily immediate effects.

It is not like taking a pill which will numb you. This is just the reverse. When you actually have body work, you engage a journey of healing. With any journey of healing it is all often the case that you sometimes do feel worse for a period of time before you begin to feel a lot better.

Perhaps you decide you'd like to experiment with bodywork of one type or another. Say you choose to try craniosacral work and you visit a craniosacral therapist. You explain what symptoms you have. The one activity you would like to manifest to walk gracefully on your daily walks in your neighborhood.

The craniosacral therapist works for a period of 45 minutes to an hour as they invite your body to release tension. It could be when you return home that day; you may find that taking a walk is actually a bit more of a challenge than it was the day before. You may be thinking to yourself

> *Wait a minute, Wait a minute. Is this the right therapy for me to pursue? I don't know whether this is working for me and not.*

If you are committed to releasing tension and trauma from your body, patience is needed. Allow whatever feelings are present to emerge so that they can essentially be released and forgotten. The trauma has likely been lingering around the tissues of your body for a long time. It may take several sessions to release it, but as the trauma is released you will be able to walk more fluidly on your walks, little by little, day by day.

- *It does take persistence.*
- *It does take a willingness to acknowledge that a journey of healing will be rocky.*
- *You will have periods when you are not feeling better.*

If your aspiration is to feel significantly better every day, look out. There will be some days when you're not going to feel significantly better. In fact there may be some days when you feel worse.

In the context of the big picture, symptoms that are flopping all over the place is what will happen when you are on the road to getting well and recovering from the symptoms of Parkinson's. The Road to Recovery is a rocky process to be experienced.

When traumas are rooted emotionally, as is the case with many physical traumas, the emotions that are connect to the traumas can be unpleasant to feel. But, it is healing for the emotions to surface. When you can actually feel through all that, the trauma is released. As it is released, the tissues in the body clear out all the

wrapping in
the tightness
the tension

that is being held in your body. Cells can now nourish themselves. Neural networks can rejuvenate themselves. Neurons can come back on line and function as originally designed.

Persistence is important. Am I being persistent enough in hammering the importance of persistence? I hope so.

When you are confronted with information that you're not doing as well today as yesterday, two responses are possible, One is:

> *"I give up. It must be true that I'm going to always get worse."*

The second response is to persist and say

> *"I'll never give up. I'll never give up. Over my dead body will I ever, ever give up."*

Persistence then is really a key component of the recovery process. It doesn't matter what happens in your decisions for what therapies or modalities or approaches that you decide to pursue. If one doesn't work, you can always shift over to something else. Stick with it. You can always find something that works.

How Can I Really Know What is Helping If Sometimes I Get Worse?

The challenge of deciding what works presents a rather interesting challenge for everyone. Let's say you've elected to experiment with a particular therapy. You get the therapy and afterward you feel worse. The question becomes:

> *"Wait a minute. Should I persist with this therapy or is this therapy actually doing me a disservice. Should I continue with it or abandon it?"*

Confusion about what to decide is typical. Is it a better choice to switch therapies because the therapy that you are currently pursuing for yourself seems to not be working well? Or, is the better choice to continue with the therapy that you've started? It can be a formidable challenge to sort through these choices, especially when you are not feeling well.

Self Administered Muscle Testing

Muscle testing is one way of asking your body to help you decide. The idea behind muscle testing is to get out of your rational mind and access the wisdom of the body. The method I will explain can be done anytime and anywhere, moment by moment, and requires no assistance and no help. It is easy to do, entertaining and quick.

You may have prior experience with muscle testing, but I seriously doubt the method I am about to explain will be familiar to you. There are many intricate and complicated ways of muscle testing. There are numerous ways of muscle testing that require a second person to help.

For example, in one of the procedures, you hold out your arm and the person helping you pushes your arm down after you ask a very specific question. Let's say you have had seven craniosacral treatments and are undecided about whether to continue. You ask the following question:

"Is craniosacral therapy helping me recover?"

You place either arm (right or left) straight out from your body so that it is level to the ground. Then, what they can do is to say out loud:
"Is craniosacral therapy helping me recover?"

Your friend now pushes your arm down toward the ground. If there is a great deal of resistance when your friend pushes your arm down to the ground, it means the body is saying,

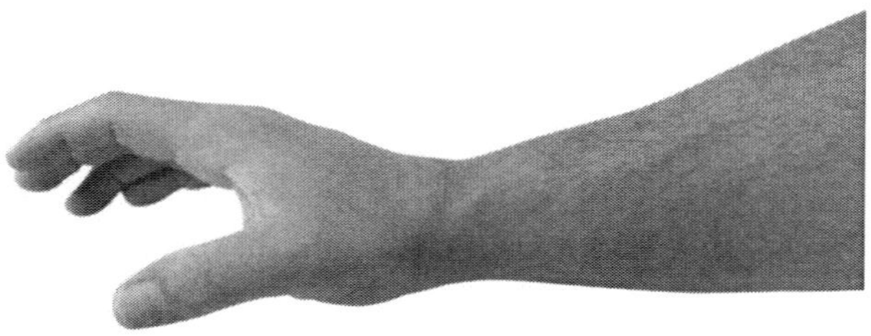

"Yes. The craniosacral therapy is helping me recover."

Your mind may have doubts, but your body does not. Stick with the therapy for now. After 10 treatments, you may get a different answer, but for now the answer is yes.

If your arm gives little resistance and flops toward the floor like a wet noodle your body is saying

"No. Craniosacral therapy is not helping me recover"

You need to look for another approach or therapy. This therapy at this time is not what your body needs. Of course, if you ask again in a month, you may get a different answer.

This is one of the ways to muscle test your body. There are many other methods of muscle testing that are also effective. Regardless of the method you use, the advantage of muscle testing is that you short circuit the clutter of your mind.

There are also ways of muscle testing that do not require the assistance of another person. We can muscle test ourselves.

Let me now explain one way of muscle testing that does not require help from another person. This particular method is cool because you can do around anyone in any situation at any time of the day or night. People will not even know that you're doing it.

This method requires the use of two fingers and one thumb from either your right hand or your left hand. You can use both hands if you want, but one hand is really all that's necessary. You always have your thumb and fingers with you, so you always have what you need to muscle test yourself when you use this method.

1. Take your thumb, your pointing finger and your middle finger (the two fingers that are closest to your thumb).
2. Rub your thumb over those two fingers in circles.
3. Take notice of the feeling. Is it smooth or is it sticky?

If the surface tension gives little resistance and feels smooth, your body is saying "Yes." If the surface tension is sticky, your body is saying "no" to whatever question you have posed.

Here is how to become acquainted with how it feels when it is sticky. Find a substance that you know is harmful to you and put it in your other hand. A good substance to use is a battery which is full of toxic chemicals. Any toxic substance will do.

Or use a food that you know you are allergic to. Simply take that food - in a package or in its raw form – or a battery and place it in one hand. By holding the substance, ask your body

"Is this substance good for me?"

Now, muscle test the answer using your other hand. Circle your thumb across your middle and pointing finger and check to see whether not there is any stickiness present.

If the substance you are holding in your hand is not good for your body, it will feel sticky and wet when you rub your thumb and fingers together. This tells you the substance is bad for your body (which of course you already know).

By taking anything that you know is not helpful to your body and putting it in your hand you can get a real sense of how sticky it can feel when you muscle test with the other hand. The feeling is quite incredible I might add. The resistance will feel clammy. It feels sticky whenever you are addressing something that's not good for your body.

Now place a food in one hand that you know your body loves. Muscle test with the other hand. Ask the question,

"Is this food good for me to eat?"

When you muscle test it will feel smooth. This means your body is telling you

"Yes, I like this food. This food is good for me"

With just a little practice you'll be able to feel the difference. Once you get a sense of what that difference is, you can muscle test whether or not a particular therapy is a good therapy for you or not.

What follows is a meditation that will invite you to use muscle testing on yourself to find an answer you have been seeking.

The Five Steps that are Necessary to Recover

People who are feeling better experiment. They try new therapies. They find ways to feel better and they are feeling better. It is not easy to be sure, but it is possible to recover as long as you can be attentive and disciplined and do it every single day. It will happen.

The five steps are the core principles of manifesting anything, whether it is manifesting:

More money
A partner
Better health

I have applied the steps to recovering from the symptoms of Parkinson's. There is no mystery that these steps work. They do work. They will succeed. It is not easy. It takes commitment. It takes discipline. It takes the commitment. Once you enter the road to recovery, you can make it happen.

Five Steps to Recovery

Focus. A concentrated focus of your thoughts is essential. When you are wishy-washy, when you flop from one desire or one want to another, recovery will flop.

Our thoughts become materialized in physical form only when we have a concentration of the same thought. If we juggle many different thoughts at the same time, the power to materialize anything fizzles. Focus of thought is essential.

Detail it Out. Once you have set in motion what it is that you most desire, the second step is to detail that out. Make sure that it becomes absolutely clear in every respect. Know precisely what it will:

Feel like
Look like
Hear like

Detail out everything you want to accomplish for yourself. If your desire is to move fluidly then be very clear for yourself about exactly what you want that to look like.

- *Do you want to be able to run marathons?*
- *Do you want to walk like a ballerina dancer?*
- *Do you want to move around the floor effortlessly as you play with*

    - *your children*
    - *your neighbor*
    - *your grandchildren*

See it the way you would like it to be and it will become so. It is critical that you not focus on what you do not want. No one ever got anywhere with a double negative. Focus on what you do want.

Stay on Track. The third step of recovery is to stay on track, to stay with it, to not waver off the course that you have set for yourself. If you have set your intention to move with fluidity, do not waiver from that intention. Do not switch midstream into another desire or another thought or another wish. Stay on track with

One dream
One want
One wish

Keep the Faith. The fourth step – and this is often a challenge for people –is to keep the faith that it will happen.

If you are now struggling with breathing problems or movement problems, if it's difficult for you to stand up or sit down or get in a car or walk then you may be thinking

Five Steps to Recovery

"Well here it is. Here it is to stay."

It's critical to keep the faith.

People with Parkinsons are recovering
People with Parkinsons are becoming more fluid
People with Parkinsons are beginning to walk with ease.

People are doing things that are making it possible for their bodies to heal. The mission of Parkinson's Recovery is to document the stories of people who are recovering, to identify what people are doing that is making a difference.

You can hear the interviews. You can hear their stories. They are not getting worse. They are getting better. It is important to keep the faith.

Persist. Be persistent. Stick with it. Never give up. We all get discouraged. If you make your personalized recovery program a daily discipline and a daily habit, you will feel better. It will happen for you. Count on it.

Over my dead body will I ever give up.
Over my dead body will I ever give up.

It is not easy to be sure, but it is possible to recover as long as you can be attentive and disciplined and do it every single day.

Count on it.

 Robert Rodgers, Ph.D.

Meditations

Persist Overview

I am so glad that you have finally arrived at the fifth and final step to being able to recover from the symptoms of Parkinson's.

You have persisted.

There are those who have started listening to the five steps and for one reason or another stopped listening

- *at the second step or*
- *perhaps the third or even*
- *perhaps the fourth*

It was, you see, just too difficult for them to persist.

What then is the fifth step to recovery? You guessed it. It is to persist.

1. **The first step is to focus.**
2. **The second is to materialize.**
3. **The third is to stay on track.**
4. **The fourth is to keep the faith**
5. **The fifth and the crowning step is to persist.**

It doesn't mean these steps have to be taken one at a time or in sequence. All of the five steps combined together will assure you of being able to find relief from symptoms in time, perhaps not in the time you had hoped for, but in good time.

This meditation today is all about strengthening your already strong sense of persistence. It is strong because you have now arrived at the final step.

So, settle in. Relax. Find a comfortable place where ever that might be

- *Sitting*
- *Lying down*
- *Leaning against a wall*
- *Sitting in a car*

Where ever you might be. Just get cozy comfortable now.

- *Invite yourself to take a breath.*
- *Focus in on the air as it brushes against the back of your throat and soothes the inner part of your lungs*

- *Focus on the air as it exits your body into the ethers of the universe*
- *Breath in.*
- *Breath out.*
- *Focus your attention only on your breath*
  and as you do bring your attention to whatever place in your body holds tension where ever you feel

    - tightness or
    - stress or
    - anxiety

and simply invite that

    - tension or
    - stress or
    - anxiety

in your body to dissipate now.

No need to force.
Simply invite.
Sink into the relaxed place you are at now into this delicious state of

Calmness
Peace
Security

- *Close your eyes.*

Five Steps to Recovery

- *Focus your attention on your breath.*
- *Relax all places in your body that hold tension.*
- *Invite your body to balance all of the many, many, many hormones that are inter-related and bound together in the same family.*
- *Breath in.*
- *Breath out.*

Today you will be taking a trip on a train. This particular train must travel on tracks over the top of a very, very, very tall mountain, a mountain that is as tall as you have ever seen in your life.

You are now at the bottom of the mountain. Your fondest wish is to get to the other side of the mountain.

1. That is the goal.
2. It is clear to you.
3. You are now focused on that goal.

In your mind's eye you have materialized what the other side looks like and it is delicious. No worry about staying on track because there really is only one track from the one side to the other side of the mountain.

It is important to keep the faith that you'll successfully be able to travel all the way up and over the mountain to the other side. Of course, what counts here today is your persistence.

You are at this point the only one taking this train trip. Place on your train everything that you need to take with you that you anticipate needing on the other side of the mountain, for it is unclear to you if you will ever be able to travel back.

Maybe so
Maybe not

To be safe you need to make sure and carry on this train all of the baggage that you believe is going to be essential for survival on the other side of the mountain. In your imagination collect together

All of your belongings
All of your essentials

Take everything that is needed aside from people for you cannot take people on this trip. Gather together everything that is essential to have when you arrive and begin living on the other side of the mountain. Do that now in your imagination.

Make sure that you place everything that will be essential on the train. There is plenty of room.

- *If you need to add cars, no problem, so please take them.*
- *Perhaps furniture is needed.*
- *Perhaps vehicles.*
- *Take what really matters.*

Make sure that you place on your train whatever is essential and necessary for you when you arrive on the other side of the mountain. Do that now in your imagination.

Make a quick mental check to make sure that you haven't forgotten anything

- *Passport*
- *Toothbrush*
- *Camera*

Remember everything you might need. Make sure that you put it on the train because after it starts up the mountain, there's no turning back.

It is now time for the train to depart from the home station. Say goodbye to whomever you need to say goodbye to. This is a journey only for you today.

Please walk up the first four steps of the train. As you step onto the fifth and final landing the door is opened for you by the conductor who is there to assist in any way necessary. Find a comfortable seat knowing that all of your belongings are safe.

Now, the train departs. Feel free to say goodbye to your friends and family and loved ones, all of those who were waving you goodbye and wishing you well on your journey, hoping to see you soon as you are hoping to see them soon.

This train is a very special train because it is a steam engine. It uses very natural power of sorts and so you can hear the chug-chug of the engine and you can even smell the steam as this train begins to zigzag

- *from left to right*
- *from left to right*

across very, very tall forests, slowly chugging its way up the mountain.

The engineer of this train - who is very skilled in every respect - pumps the boiler with more and more wood so that there's plenty of power for this steam engine to take you and all of your essential belongings all the way up to the top and over to the other side.

- *Smell the steam*
- *Hear the clank of the wheels running over the tracks*
- *Watch the deer and the squirrels and the beaver who are collecting as beavers do materials for a dam over a most gorgeous green colored lake*
- *Look at the glitter on this lake and the seductive nature of the color of this water which drains from the glaciers that live higher up the mountain.*

You have now reached halfway up the mountain. What a divine sight this is as you chug-chug along up the mountain higher and higher and higher.

- *Push a window down and listen to the birds.*
- *Listen to the sound of the coyote.*
- *Look at the marks on the tree which are the marks of a bear that has crawled up this tree to the top so that she could have a good vantage point.*

The train shrugs up the mountain little by little, going a bit slower now for the mountain is getting steeper and the train is beginning to struggle. You can tell that by the slowing down of the sounds of the wheels clanking against the rails. There is a bit of swaying back and forth on this train.

The trees are getting smaller and smaller. There is less foliage and fewer animals now for the forest is getting tinier and tinier. Even the number of birds are becoming fewer and fewer.

I hope you have brought a sweater along for it is beginning to get a bit chilly up at this altitude. Feel free to clothe yourself in warm clothes.

The train continues to go up this very, very steep mountain. Now it looks like you can almost see the top - not quite – almost. Somewhat alarming is that this particular train seems to be going

- *slower and*
- *slower and*
- *slower*

as the mountain gets

- *steeper and*
- *steeper and*
- *steeper.*

Suddenly, in the front of the train car, a door opens and what appears is a man you have not seen before. This man must be the engineer. He wears the hat of an engineer and the clothes of an engineer.

The man rushes somewhat frantically to where you are sitting for remember you are the only one on this train. He says to you

> *"I have something to report of some concern. I thought that enough wood had been loaded on to the engine but for some reason - and I cannot tell you what it is - we don't appear to have enough wood for the train to actually get to the very tippy top of the mountain and to the other side.*
>
> *"No problem if we can just get to the top for as you can guess, we will be able to just slide down the other side. The brakes work well enough.*
>
> *"I don't know really what to do about it. I'm quite frantic. I do, I do apologize to you because this is certainly not what I thought what happen for you on this (what I know) is a most special and important journey that you are taking.*
>
> *"But, I might suggest - if I could – there is something you could do that would help greatly I think. You see, if we can just shed some weight I believe there's a chance that we could actually pass over the top of the mountain.*
>
> *"I know this is a lot to ask but it is -- it is really to the point of an "if" or "if not". That is, if you could just take some of what you have put on your train and if you could just toss it off the*

side we'd have less weight. Then I do think - I do think - that will be able to get over the top.

The engineer turns and moves to the very front of the car, opens up the door quickly and exits in a flash because, you see, his work is in the engine car.

There is one choice. Please begin throwing overboard whatever you can find, especially those heavy items, for the engineer must be right. This is the only way to get over the top of the mountain. Find whatever you can find to toss.

Did you bring a wheelchair? You are going to have to toss it over. That is certainly heavy.

Did you bring a metal cane? Sorry but that's got to go that has too much weight to it to be able to be of any benefit now.

What did you bring on your trip that has a lot of weight?

Did you bring notifications about nursing homes - all of those brochures you know about places that you can go when you get older? Did you bring those brochures in those materials with you? I'm sorry those must be very heavy. Those must be tossed overboard now for this train will never go over the top unless there is less weight there.

Quickly – toss whatever bags you have brought that have weight and that really you can live without. Please toss them over now for this train will never make it over to the top unless you do.

There is no schedule for a second train. This is it. This is your chance. Either you do it or you don't do it.

Scan through everything that you've brought on the train. In the short time that you have left,

- *Toss stuff out the windows.*
- *Toss it out the doors.*
- *Toss out everything you can put your hands on.*
- *Get rid of every item you can possibly get rid of that you have brought with you.*
- *Now.*
- *Everything.*

A popping sound snaps again at the very front of your car. The engineer's face is again in your face.

> *" I'm - I'm very sorry. I know what you are doing is helping, but it doesn't look like we're going to make it. Can you please - anything else - please throw it out. Not yourself of course. You must stay. Is there anything else you can throw overboard? If it is not essential, please - please discard it.*

The engineer flies away in a flash the door slams.

What else is there to pitch?
What else is there to pitch?

There must be something else that can lighten up this train. Ah, I know something you forgotten about.

Collect together now each and every one of the negative thoughts that you have brought with you on this trip. Do it now, all of them. Collect them all together, for those must also go overboard if you expect to get to the top of this mountain.

- *Collect them together*
- *All the negative thoughts*
- *All the judgments*
- *All the negatives*
- *All the reasons why not*
- *Collect them all together*
- *Now everything*
- *Every negative thought*
- *Throw them overboard*
- *Get rid of them now*

I can see it. I have to tell you. There's one negative thought that you're still holding. You haven't tossed it overboard yet. It is going to be necessary for you to toss that one overboard if you expect to get over the top of the mountain.

That's the thought that you're not to make it. Toss that one.

Oh is there another thought. It is the thought

"I might as well give up

> *This is too much trouble*
> *I have given up too much already"*

Do you have these negative thoughts? If you do, it is time to toss them out too. Do not hesitate. Time is running out. You are not going to make it over the top of the mountain if you hold onto those thoughts.

There are other thoughts that do help

- *I can do it*
- *I can do it*
- *We can do it*
- *We can do it*
- *Over my dead body will I ever give up*

- *I can do it*
- *I can do it*
- *We can do it*
- *We can do it*
- *Over my dead body will I ever give up*
- *This will happen*

- *I can do it*
- *I can do it*
- *We can do it*
- *We can do it*
- *Over my dead body will I ever give up*
- *This will happen*

The train continues to struggle. The clinks of the wheels on the tracks are now very few and far between. You can see the top of the mountain but you are not quite sure how far it is. There must be at least 100 yards left to be able to get to the summit, to the place that's required before you can gently slink your way down the other side of the mountain.

Hear the clank of the door at the front of the car again. The engineer is again in your face and you're a bit sick of looking at him.

> *"So, here's the deal. I'm out of wood. No more fuel for this engine. As you can see, we have no more trees to chop. Have you tossed everything?*

I'm afraid the answer is yes. You get the flash of a crazy idea.

> *"I know what to do. How about pushing the train? Would that help?*

> *"Oh,*

the train engineer says,

> *"I've never heard of such a thing. We've never done it that way. I've always gotten over the mountain for other people. There's always been enough fuel. I'm sorry. This has never, never been the case for me. I have never had to ask anybody to get out of the train and push the train over the mountain, but it looks like it's the only choice left. Well,*

the engineer says

"why don't we give a try?.

The train now is chugging up to the top of this mountain very, very slowly - almost about to run out of fuel. You and the engineer exit very carefully out the steps of the train and wait for the rear of the train to approach. Now,

- *Push*
- *Push*

with whatever power and strength you have.

- *Push this train*
- *Inch by inch*

so that it will reach the top of the mountain. Set your intention to go into a place of martial power - the place is called tan tien. It is two inches below the belly button. Fire that spot up red engine hot.

You might have to help the engineer do the same. He's not in that great physical shape. Use martial arts strength. In this place you have the force and the power to push anything

Anywhere,
Anytime
Of any weight

Five Steps to Recovery

- *Take a breath*
- *Get set.*
- *Focus.*
- *Materialize the train at the very top of the mountain*
- *Keep the faith that it will happen*
- *Push*
- *Push now*
- *Push the train*
- *Push*

It is now inches before the top of the mountain.

- *Are you ready to give up*
- *Can you feel that energy?*
- *Are you ready to give up?*

Feel that energy now. It has a low frequency. It will get you nowhere.

Turn the switch. Turn the switch on.

- *I can do it*
- *I can do it*
- *Over my dead body will I ever give up*

Reset that martial power now. Bubble up that tan tien. Make it red hot. Help the engineer do the same.

On the count of three push once again one to get set for push to train up together step-by-step as now the train begins to go

Five Steps to Recovery

A little bit faster
a little bit faster
just a few inches to go
almost to the top
getting set
Feeling the power of persistence

The engineer all of a sudden runs over to the side jumps into the train because, you see, somebody has to be in the engine for the train might fall all the way down the other side.

- *Push.*
- *You are the only one pushing now.*
- *Push*

It is only up to you for this train to be at the very top.

- *Push.*
- *Feel the power of persistence*
- *Feel what it feels like to make it happen*
- *Focus.*
- *Materialize what it will feel like when you're at the top of the mountain.*
- *You are staying on track*
- *You have the faith that it will happen*
- *It will happen*
- *It will happen*
- *Give up over my dead body*

- *Push*

By God, what you now notice is that the train all of a sudden is picking up speed. It is beginning to go over on to the other side of the mountain. The engineer stops the train because, of course, you must get back on.

Before you do, walk very, very

slowly and
proudly and
competently and
powerfully

to the summit of this mountain.

- Stand now on the summit of this mountain. Look down to where you have been before.

- See all of the strewn baggage on the mountainside the baggage that you no longer need to carry with you to the other side.

- Look over at your train, the little train that could

- Take your two arms and raise them high up into the air as you stand on the top of this mountain.

- Celebrate the victory of persistence

- Celebrate and honor the power of the energy behind what it takes to manifest your heart's desire

    - *You have made it*
    - *You have arrived*
    - *You are there*
    - *Celebrate now*
    - *Feel the sweetness of mother earth under you and the air as the tangles across the back of your throat.*
    - *At last you have arrived at the summit, the place where your heart's desire sought to be today.*

- *Look now at the other side of the mountain to the place where you are about to go effortlessly.*
- *Look down as best you can to where you are going*
- *In your mind's imagination see what the other side looks like*
- *Where are you going?*

With focus, materialize what the other side of this mountain looks for you. It's easy to keep the faith now because this train needs no more power to effortlessly get you where you are going.

Your persistence has paid off
Hallelujah! Hallelujah!
I have arrived at last

Carry with you always with each and every challenge you give to yourself the energy of this meditation.

Five Steps to Recovery

- *I can do it*
- *I can do it*
- *Over my dead body*
- *I will never give up*

Persistence will pay you handsome rewards.
You can count on it

Should I Persist or Switch?

- Get settled.
- Get set.
- Get relaxed.
- Take a breath in and a breath out
- Relax whatever tension might be present now
- Set the intention to ask your body for the answers.

Your body always knows what it needs. You just have to remember to ask. The answer can change day by day. What the body needs today may be different from what the body needs tomorrow. Muscle testing for the answers can be so helpful. Once we get the answers, we know

When we need to persist
When we need to switch

There are two parts of persistence that need to be acknowledged. The first part of persistence is experimenting, taking positive action and trying a new therapy. It involves sifting through the many choices. It requires making a choice about what you can do to feel better.

The second part of the persistence is to evaluate the therapy once it has been selected. Should you stick with it or find another therapy (or another therapist)? You can use this method to ask about any therapy, be it a medicine, exercise, food, supplement, counseling, body work – you name it.

- *Take one more breath.*
- *Just relax now.*
- *Settle into your chair or wherever you might be.*
- *Focus your attention on your breath with attention to how it leaves your body.*
- *Invite yourself just to settle in now.*

Take your thumb and your two fingers and have them handy. I know you have brought them with you. I am going to ask your body some questions and you will need your thumb and fingers to learn the answers.

Think back now - either recently or over the last year or two - about a therapy that you decided to pursue.

- *The therapy might be a prescription medication*
- *The therapy might be an exercise routine*
- *The therapy might be a psychologist or a counselor that you have decided to see*
- *The therapy might be body work like Bowen therapy or craniosacral therapy*
- *The therapy might be exercise of a certain character or nature*
- *The therapy might be eating a certain type of food that you were told should help*
- *The therapy might be taking a specific supplement that was recommended by an internist or a friend*

Think back on a therapy - whatever that therapy might have been - that you still have questions about. You find yourself wondering,

"Should I still be doing that?"

Whether or not you decided in the past to continue or discontinue the therapy, focus on that particular therapy now. Once you have that therapy in mind, make the statement:

Doing [insert the name of whatever therapy you have decided to ask about here] is a good choice for my body."

Now, muscle test. Take your thumb. Rub it against your two fingers and see how that feels. Do that right now.

Make the statement again as a double check.

"Doing my [ ] is a good choice for my body. I need to continue doing this."

Again, take your thumb and your two fingers and rub them together. Be sensitive to how it feels. Is it smooth or is it clammy?

If it feels smooth it's a great idea for you and you need to continue. If it feels clammy, it's not a good idea for you and it would be important to consider discontinuing it and looking for another therapy or perhaps a complementary therapy.

You can now do this for a second therapy of choice, whatever that might be, something that you really couldn't figure out whether it helped, whether it was

- *Eating or*
- *Exercise or*
- *Going to a counselor or*
- *Going to bodywork or*
- *Detox program*

What thoughts do you have about whether or not that therapy was a good idea or bad idea? Go ahead and ask the question as it comes to mind:

- *Is doing [fill in the therapy] or*
- *is taking [fill in the therapy] or*
- *is trying [fill in the therapy] what my body needs now?*

You can ask that out loud or with your thoughts. Then, simply muscle test by rubbing your thumb over your two closest fingers. See how it feels.

Trust that your body will give you the information. Your body always knows what the answer is. This is another clever, powerful way of being able to get information from your body about what it needs.

We get ourselves out of our minds and connect with the wisdom of the body when we muscle test.

This way of muscle testing is no big hurdle. It's no big deal. We can always muscle test any question in this way hundreds of times a day if we want.

The answer may be in between yes and no. The feeling may be part clammy, part smooth. If the feeling is not totally clammy and sticky but somewhere between sticky and smooth, then it may be the body could benefit from the therapy but it is not necessarily the best thing for your body right now. There can be an in between answer that your body will give you sometimes when you do this particular test.

Try it out for size. When you get the answer

"Yes. My body needs this therapy."

Activate that precious fifth step that is so essential to manifesting recovery. Persist, even if in your mind, you think the therapy is making you feel worse. If your body says:

> **I need this.**
> **I need this.**

Then your body does need the therapy. But keep asking, the answer can change.

There are many different opportunities and possibilities out there that are working for people. The number keeps growing every week. I am adding therapies to the list of good possibilities every week. Begin communicating with your body so that you can figure out what therapies and what modalities will be most useful for you.

To summarize:

- *Muscle testing is a clever way of getting out of the way of your mind.*
- *Muscle testing is a clever way of listening to the wisdom of your body.*
- *Muscle testing is a clever way of connecting with what your body actually needs.*
- *Muscle testing is a clever way of honoring the critical role of persistence.*

Finish What you Start

How many times in your life have you given up?

- *Started at training program and never finished*
- *Started a knitting project and put it aside*
- *Began a cooking program and decided it was too much trouble*
- *Started a job and quit before you even began*

Think back over your life from the time you were little until today. How many times have you been a quitter? How many times have you given up when you didn't want to give up? It was just too much trouble to continue.

There were other things that were calling, but in your heart you really did want to finish.

Of all those experiences when you were a quitter, let one float to the top now - one experience that you gave up - that you quit - which you find yourself regretting. Let one float to the top, one that you regret quitting.

- *What is it?*
- *What do you regret quitting?*
- *Why not rekindle this interest?*
- *Why not do it now?*
- *Why not re-activate the energy that you used to start this activity?*

No excuses allowed. It is not a question of being ill or full of symptoms or old.

If it is in your heart's desire to finish this activity,
then get ready to finish it.
Now.

The Power of Persistence

Remember a time in your life when you gave up. Remember a time. Whatever comes to consciousness is the memory that you need to work with today.

Go into that place and to that space. Feel what it feels like at the moment you decided

"I'm not doing this anymore."

What is it feel like? What is it feel like in your body? Remember the moment. Remember where you were and what you said to yourself or to others and capture the feeling in your body.

Now, remember a time in your life when you persisted when you refused to give up. Allow whatever memory float to the surface of your consciousness that you need to work with today -

A time when you get it
A time when you made it happen.

Remember the moment when that happened. Revisit that time and place and that experience. Remember what it feels like in your body.

With your intention notice the difference. Notice the difference between what it feels like in your body when you complete something you started, when you persist despite all obstacles and problems with what it feels like in your body when you give up before something is complete and finished.

Notice the difference.

Notice what it does to you physically.
The difference is huge.

The energy behind persistence is all powerful. The energy behind the "give it up" place is powerless. It gets you nowhere.

- ***Feel the difference.***
- ***It is huge.***

The choice is yours.

Persist or give up.

If you want limitless energy then persist. If you want to be powerless, then choose to give up.

Persist or Quit: It is a Choice

As you have worked over the five steps to recovery what has floated to the top for you in terms of your intention, in terms of the wish that you would like to see fulfilled? What is it? What have you decided to focus on? Bring that now to your attention.

This is what you have focused on. This is what you have been materializing in detail. You have stayed on track and you have kept the faith that in fact this will happen. Bring it to mind.

Now, think about what obstacle you will confront or have confronted. Bring it to mind. Go ahead and see it.

- *It is an obstacle.*
- *It is a problem.*
- *It is a hurdle.*

Go ahead and acknowledge it. What is your reaction? Are you ready to give up? Are you ready to push through all of that and move to the next place?

What is your reaction? It is a choice.

Persist or give up

It's a choice.

- *More obstacles*
- *More problems*
- *More difficulties*

You've done a lot. You've tried a lot. It's not working.
Are you ready to give up?

Want to be a quitter?

It is a choice. You can give up or you can persist. Make a choice and either move to the next place, to the next step, or move backward.

Settle in. Be a couch potato. Drift away. Feel the energy-less place that this puts you in.

Or, feel the energy pulse and sensation and power of what it feels like to move through

- *Obstacle after obstacle*
- *Problem after problem*
- *Difficulty of the difficulty*
- *Setback after setback*

Feel the energy and the power even as more and more problems emerge with persistence. It is a formidable power indeed. I invite you to embrace the power of persistence now.

Where are You Looking for Answers?

Bring to consciousness again today the intention that you have been working on coupled with listening to these recordings. Bring it to mind.

I strongly suspect it's a wish that you would like to see fulfilled.
It's a wish that you are focusing on and materializing. You are staying on track because you're still with it and you have the faith that something will in fact happen with this wish.

Bring it to mind for this is what we will work with today. As you work with this, identify what problems or obstacles you have confronted in moving through this most sacred desire. If it's new you may not have encountered them so you'll have to imagine what those obstacles might be.

What are they? What have been the problems? What have you tried? Remember what they are. Acknowledge your efforts with the intent of feeling better.

It could be you have tried

- *Many therapies*
- *Many approaches*
- *Many avenues*

It could be you have talked with many people. It could be you have spent quite a few years on this. Acknowledge that.

Now, I invite you to ask another question.
That question is:

What am I missing?
What possibilities have I not considered?
I must be digging a hole in the wrong backyard.

Go to another backyard. Start digging another hole. Start looking in another place.

There has to be an answer out there for you somewhere. Talk to new people. Investigate literature you haven't seen before. There are countless options and opportunities out there in the world. Why limit yourself?

Investigate.
Research.
Persist.

Five Steps to Recovery

If the hole you're digging in your neighbor's yard is also not yielding any beneficial results, go elsewhere. Look elsewhere.

Persist
Keep looking
Keep looking
Keep investigating
Keep experimenting
Keep trying out possibilities.

There is an answer for you. Give up only over your dead body.

It will happen.
It will happen.
Persistence will pay off.
It will happen for you.
It will.

Put Reminders in Your Calendar

The meditation today is an assignment. Get out your planning calendar, whatever form that happens to take for you. Get it out now. Turn this off and then turn it back on what you have your calendar in hand. I'll wait.

Great are you back now?

I'd like to invite you to first of all notice what the date is today. Then skip across one week ahead in your calendar and write the question on that day:

1. Am I focused?

Write this question in your calendar one week from now right now.

Two weeks from today in your calendar write the question:

2. Am I materializing?

Remember that this is Step Two of the recovery process. Are you detailing out what it is that your heart's content wishes to see happen? It's not a one-shot process. It's a continuous process day in and day out. That's what gives it power. Write this question in your calendar two weeks from today.

Three weeks from today write the question:

3. Am I staying on track?

Just a question to write in your calendar to ask yourself three weeks from today: Am I staying on track?
Four weeks from today write the question:

4. Have I kept the faith?

Just write it in your calendar, a question you will see on that day when you look at your planning for the week.

Four weeks from today, write the question

5. Have I persisted or did I give up?

Write that in your calendar. Now. Do it for yourself.

Five weeks from today there is a final question I invite you to write in your calendar.

Do I need to listen to these recordings that discuss the five steps to recovery again?

Write that question in your calendar five weeks from today. Just write the question. You will see it again in five weeks.

Recovery is a process. It is not an event. It unfolds. It's a lifetime work. It doesn't just get turned on and turned off. Make it a process and you will be able to manifest your heart's desire.

Many, many blessings and may your recovery unfold now.

Pioneers of Recovery Discount Coupon Code

Thanks for your interest in the work of Parkinsons Recovery. A companion book to Five Steps to Recovery is Pioneers of Recovery which was also released in 2009. The 364 page book contains information about what seven pioneers with Parkinson's report helps them get great relief from their symptoms. I also interview experts in various fields (herbalist, aroma therapist, Neuro-linguistic programming practitioner, psychologist, physical therapist, comic) who discuss their experience with what can make a huge difference to how you feel.

Claim the print version of the book at a 25% discount. Here is how you claim the discount.

1. Visit Pioneers of Recovery.

2. Click on the shopping cart link to claim you're the print book. Regular price of the book is $59.95.

3. On the shopping cart page, enter the coupon code "relief" (without the quotes). The price will immediately be adjusted to show the 25% discount.

Index